FI

TRANSDUCER INTERFACING

Signal Conditioning
for Process Control

TRANSDUCER INTERFACING

Signal Conditioning for Process Control

Robert G. Seippel

A RESTON BOOK

PRENTICE HALL, Englewood Cliffs, New Jersey 07632

Library of Congress Cataloging-in-Publication Data
Seippel, Robert G.
    Transducer interfacing: Signal conditioning for process control
  Robert G. Seippel.
      p. cm.
    ''A Reston book.''
    Includes index.
    ISBN 0–13–928888–0
    1. Process control—Data processing.   2. Interface circuits.
  3. Transducers.   I. Title.
  TS156.8.S44   1988
  629.8—dc19   87–26472

Editorial/production supervision and
interior design: Margaret Lepera
Cover design: 20/20 Services, Inc.
Manufacturing buyer: Lorraine Fumoso and Peter Havens

Printed in the United States of America

10   9   8   7   6   5   4   3   2   1

ISBN   0-13-928888-0   025

PRENTICE-HALL INTERNATIONAL (UK) LIMITED, *London*
PRENTICE-HALL OF AUSTRALIA PTY. LIMITED, *Sydney*
PRENTICE-HALL CANADA INC., *Toronto*
PRENTICE-HALL HISPANOAMERICANA, S.A., *Mexico*
PRENTICE-HALL OF INDIA PRIVATE LIMITED, *New Delhi*
PRENTICE-HALL OF JAPAN, INC., *Tokyo*
PRENTICE-HALL OF SOUTHEAST ASIA PTE. LTD., *Singapore*
EDITORA PRENTICE-HALL DO BRASIL, LTDA., *Rio de Janeiro*

Contents

4—COMPONENTS OF ANALOG SIGNAL CONDITIONING   52

5—COMPONENTS OF DIGITAL SIGNAL CONDITIONING   88

6—TRANSDUCER SIGNAL CONDITIONING   108

7—CONDITIONING FOR CONTROL DRIVERS   158

8—MICROPROCESSOR CONTROL   175

APPENDIX   COMMON CIRCUITS FOR BLOCK DIAGRAMS
IN TEXT   187

INDEX   199

Preface

In today's world, it is necessary to measure everything; all mechanical things, all the toys we play with, and all the functions involved with communication are carefully monitored. We are conscious of every parameter of the weather, of the vital phenomena of the body, and all other physical happenings that prevent us from, or impede our progress as, world builders, shakers, and movers. We check the quantity and quality of everything from sunshine to earthquakes. We have also learned to live with the computer, whose task it is to keep track of all these wonderful things we have become accustomed to. The computer performs as our personal mathematician, our statistician, organizer, designer, and library. Now the microprocessor as well controls this work, in order that we may continue to live in our accustomed manner.

We are truly living in an awesome time. We have the ability to measure and control many of the things that effect our lives. Unfortunately however, physical parameters all have different characteristics. We cannot feed our computer a rain storm and expect it to control the valves of a flood control basin. The vital signs must first be measured. Measurements must then be converted to a language that the computer can understand, since the computer has limited ability to operate by external variables. All the things we measure are not always consistent, and, to make things worse, are in many different forms.

It is the purpose of this book then, to bridge the gap between the output of the transducer and the microprocessor (computer). This gap is called *signal conditioning*. Signal conditioning is the method by which a transducer's analog (variable) output is converted to a form that the computer can both understand and deal with. The first chapter of the book is a review for some readers on the fundamentals of control and control systems. Chapter two provides the essentials of measurement, including the features that have evolved with process control. Chapter three includes descriptions of transducers, sensors, and detector elements. This chapter is relevant for an understanding of the entire book. Chapters four and five provide the fundamentals of analog and digital signal conditioners. Chapter six describes the methods of conditioning signals derived from transducers to an analog level convertible to digital language, and chapter seven provides an interface between the computer and the control device. Finally, chapter eight is designed to tie chapters six and seven together, integrating the total control system. The appendix is a library of common circuits used within the signal conditioning chapters.

It has been assumed that the reader has some background in electronics exists in the reader. A knowledge of transducers and computers is also helpful for understanding. It is hoped that the experienced designer will utilize the chapters on signal conditioning as a reference and "idea bank," and that new people in the field will use the book to gain an understanding of measurement and control. The material has been presented in as simple a form as possible to fulfill the particular needs of the experienced and the aspiring.

A special thank you is given to Bob Eisenhauer, Automatic Control Systems Electronic Supervisor, Hydraulic Research Textron, Valencia, California, who provided a depth of signal conditioning expertise derived from experience. He was also gracious enough to edit the manuscript. And as usual, appreciation is given to my artist JoAnne Bline, and my word processor operator Diana Sunker, for their excellent work under pressure. A final tribute is given to the multitudes of manufacturers of transducers, and to the hundreds of engineers that I have had the privilege of working with over the years.

Robert G. Seippel

Transducer Interfacing

Signal Conditioning
for Process Control

1

Process Control Defined

Process control is a means by which a quantity of interest within a machine or mechanism is controlled, maintained, or altered. The process that is controlled may be a function of the machine or a product produced by the machine.

Obviously, the reader will have to make judgment of these statements. However, when a simple function such as the heating system in a home is suggested, not much judgment is required. The homeowner sets the thermostat for the desired temperature. When the room air temperature drops to that set point, the furnace turns on automatically. The air warms to the set point, as sensed by the thermometer. The furnace turns off at the set point. Temperature has been regulated. A simple *single-variable* control process has been defined.

We are all familiar with the automobile. The driver of an automobile determines the route and selectively modifies the speed and acceleration. Speed and acceleration are changed by stepping on the accelerator. The stronger the force, the greater the acceleration. The accelerator, through linkage, the carburetor, and all the other engine systems, controls the speed and acceleration of the automobile. The multiple variables of auto weight, road conditions, hills and valleys, and so on, are all controlled simultaneously by the operator. The speedometer and odometer monitor the speed, acceleration, and distance, and fuel quantity being consumed is indicated by the fuel gage. The cooling system stabilizes the engine temperature. The electric system coordinates power for lights, engine operation, radio, inside heater and defroster, and so on. The brake system allows slowdown and stopping. The automobile is a *multivariable control process.*

In the machining industry, manufacturing is done in stages or sets of operations that lead to an end product. In the farming industry, processes are made at different intervals, which leads to the controlled growth of food crops. Process control of farm products is not as exact as those of a machine shop, for the variables of weather are more difficult to maintain or alter.

Operating a radio is a process control. We turn the radio on, tune the dial to a station, and then fine-tune it to isolate the frequency of the signal being transmitted. The designer of the radio had in mind a common thought involving efficiency and fidelity which provides the process with the variable called quality. The process controls within the radio work toward this goal.

The concepts of the control system, process control, and processes being controlled are all intertwined. The purpose of this chapter is to separate and evaluate the principles of process control. The elements will be organized into simple, understandable segments.

To begin the chapter, a simple closed-loop system and open-loop system are illustrated. This is followed by a description of several varieties of process, including single, double, compound, and cascade processes. Let us consider a simple liquid control system.

FUNDAMENTAL CONTROL SYSTEM

Figure 1–1 illustrates a fundamental liquid control system. This system is used in large reservoirs as well as small water control areas. We have simplified the system to achieve understanding.

In this system the *process* includes the input piping, the tank, the output piping, and the liquid. The liquid level is constantly changing, and is thus called

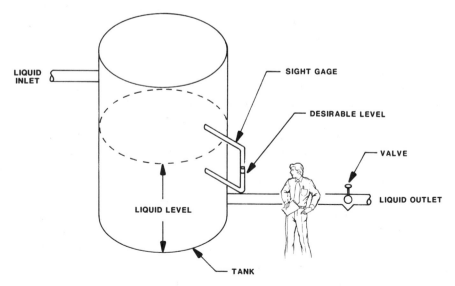

Figure 1–1 FUNDAMENTAL CONTROL SYSTEM

a *dynamic variable*. We can control this level; therefore, the variable is a controlled variable. A person is the *controller*. The valve is the control element. The sight gage is the *measurement device*. The measurement device is marked at a predetermined desirable level called a set point. The set point is reached when the level of the liquid reaches the set point.

Liquid is poured into the tank at an uncontrolled rate and amount. If we were to leave the system without control, it would probably fill to overflow. Lack of control is the case in an *unregulated* or *open-loop* system. In a regulated system we attempt to control the set point as closely as we can by *feedback*.

The controller, a person in this case, is constantly *evaluating* the relationship between the set point and the actual liquid level. When the sight gage indicates to the controller that the liquid level has risen above the set point, the person opens the valve (control element). The valve provides an outlet for the excess liquid. When the level of liquid falls below the set point in the tank, the person closes the valve.

Information from the sight gage was picked up by the controller's eyes; in turn, a *feedback* was sent to the valve via the controller to *regulate* the level of liquid. The *regulated system* is a closed-loop process. This simple single-variable process has all the functions of a more complex system. The functions for closed-loop regulation are as follows:

1. Measurement of the controlled variable
2. Comparison of the controlled variable and a set point
3. Determining the amount of difference (error) between the controlled variable and the set point
4. Directing a control element to remove the error
5. Feeding back a correction to the process to return the variable to the set point

In order that the process control be operative and continuous, these functions must be accomplished in sequence and repetitively.

Closed-Loop Control System

The block diagram in Figure 1–2 is of a closed process control loop. The components of the closed loop are the controller, the control element, the process, and measurement. This loop compares with the pictorial process control system presented in Figure 1–1.

The controller consists of an error detector and a signal processor. The error detector is simply a summing point (from electronics) of the set point and feedback signals. The output of the error detector is an error signal.

$$C_E = C_{SP} - C_{FB}$$

where

$$C_E = \text{controlled variable error signal}$$

$$C_{SP} = \text{controlled variable set point}$$

$$C_{FB} = \text{controlled variable feedback}$$

The signal processor performs the function of comparing the desired value of the dynamic variable (*set point*) with the measured value of the dynamic variable (*feedback*). The signal processor further determines what action should be taken. This action is relayed in some command form to the control element. The control element provides the drive to bring the controlled variable back to its set point. The control element causes the process to be modified. Meanwhile, the measurement element monitors this change. The monitored signal is a measurement of the controlled variable. This value is returned to the error detector as a feedback signal. In the error detector it is again compared with the set point to achieve an error signal. This process will continue until there is no error.

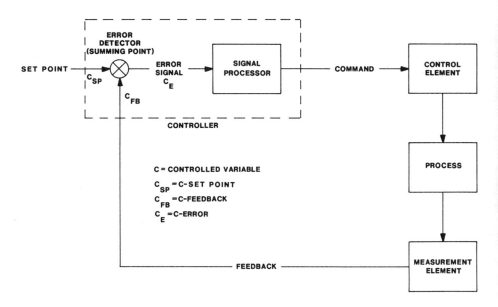

Figure 1-2   CLOSED-LOOP CONTROL SYSTEM

Open-Loop Control System

An open-loop control system is represented pictorially in Figure 1–3. The liquid level is the dynamic variable. Neither the rate and amount of flow into the tank nor the liquid output are controlled. It may be assumed that the tank will

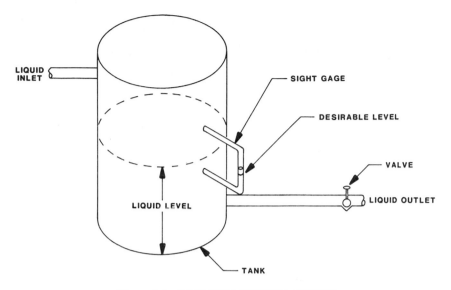

Figure 1–3　OPEN-LOOP CONTROL SYSTEM

overflow if too much liquid is placed in the input. Without feedback from an operator (controller) the process is running *open loop*. There is no regulation. The liquid level is not being maintained. An example of an open-loop system is a reservoir which overflows into a catch basin when it overfills. Open-loop systems are desirable when an output remains constant for a constant input. If external conditions remain unchanged, open loop is a feasible operation.

THE PROCESS

The word *process* means a particular method of doing something, generally involving a number of steps or operations. In a machine shop, a length of steel round stock is machined in several different operations to produce an end product. The end product will have specific uses and will conform to specifications of structure. The end product is a result of the total operative steps taken.

　　In a bakery, flour, yeast, water, and other additives are mixed, made into dough, allowed to rise, and baked under controlled conditions. The result is bread which has the properties desired by consumers.

　　Flight of an aircraft is dependent on a variety of conditions. Just to keep it aloft, several dynamic variables, including angle of attack, thrust, lift, and drag, are carefully coordinated. Variables such as altitude, airspeed, and wind velocity also come into play. Most of the variables may be regulated; some cannot be. It is also realistic to say that all the parameters of flight are interrelated and must be acted on all the time.

It is desirable, if not mandatory, to control all dynamic variables. Some are individual and must be acted on alone. These are called single-variable processes. If many elements are involved, the process is called a multivariable process.

There are a multiple of process types. The type is dependent on the job to be accomplished. Using the tank of liquid as a medium, we shall discuss, in general terms, the following processes:

1. Single-variable process
2. Two single-variable processes with interacting functions
3. Two dependent single-variable processes
4. Compound-variable process
5. Cascade process

Single-Variable Process

The single-variable process is one that has only one element to be controlled. An example of this is shown in Figure 1–4. The process under control is the level of liquid in the tank. A *liquid-level sensor* is placed in the tank and set at a desired level. As the liquid input increases, the level of the liquid in the tank increases to the set point. When the set point is reached, the controller sends instructions to open the control valve. The instructions continue until the level of the liquid has been lowered to a safe or desirable level. Continuous information from the sensor to the controller may be provided for constant level information and continuous error correction. Switching action by the sensor is also acceptable for some applications.

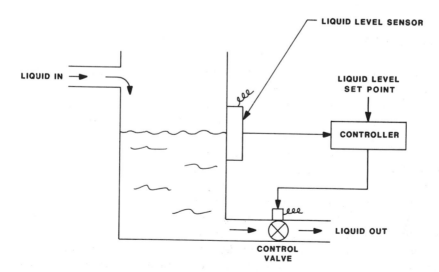

Figure 1–4   SINGLE-VARIABLE PROCESS

Two Single-Variable Processes With Interacting Functions

In this application there are two independent variables, liquid level and *liquid temperature*. Each variable is independent of the other. A complete process loop for the liquid level includes the liquid-level sensor, the control valve, and the controller (see Figure 1–5). The liquid-level sensor senses the level and informs the controller. The controller compares the information to the set point and instructs the control valve to open, thereby lowering the liquid to an acceptable level.

A complete process loop for the temperature includes the *thermocouple*, the controller, and the heater. The thermocouple senses the temperature of the liquid and informs the controller. The controller compares the information to the set point and instructs the heater to heat the liquid to a temperature that is acceptable.

These two independent processes are independent but interactive. As the level of the liquid in the tank changes, new liquid is flowing either into or out of the tank. The change in liquid will invariably cause a change in liquid temperature. Therefore, the action of one process interacts with the second process.

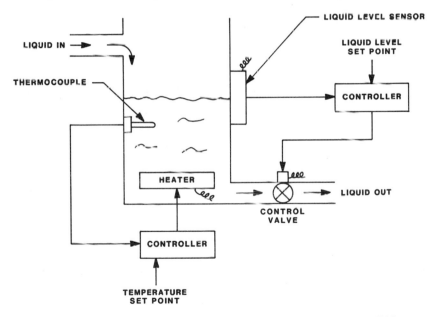

Figure 1–5   TWO SINGLE-VARIABLE PROCESSES WITH INTERACTING FUNCTIONS

Two Dependent Single-Variable Processes

The processes shown in Figure 1–6 have two liquid-level sensors. The input liquid-level sensor senses the liquid level as the measurement device that provides feedback for a flow control valve at the input. The amount of input flow is controlled

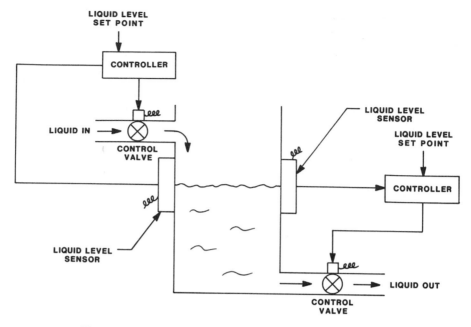

Figure 1–6 TWO DEPENDENT SINGLE-VARIABLE PROCESSES

or stopped when the level has reached the set point. The output liquid-level sensor senses the liquid level as the measurement device that provides feedback for a flow control valve at the output. The amount of output flow is varied or stopped when the liquid level has reached the set point.

Dependent operation of these control valves ensures that the level of the tank never exceeds the set point. If a failure occurs at the output control valve or, indeed, the load that valve feeds, the level sensor at the input will detect and inform the input control to stop input flow. If a failure occurs at the input control valve, the level sensor at the output will detect the level and inform the output controller to open the output control valve. This action will dump the necessary liquid to bring the tank level toward the set point. The sensors can be switches that are either on or off or can be variable sensors that provide a range with high- and low-level detection. Whichever may be the case, the processes are dependent on each other for over-level protection.

Compound-Variable Process

In the event that two or more variables are related to the same process, a single controller may be used. This action is represented in Figure 1–7 by two flow inputs to a single tank. The purpose of this process is to ensure that the two flow rates remain at a defined ratio. The method places the flow sensor X in one input line and the flow sensor Y in a second input line. Both flow sensor outputs

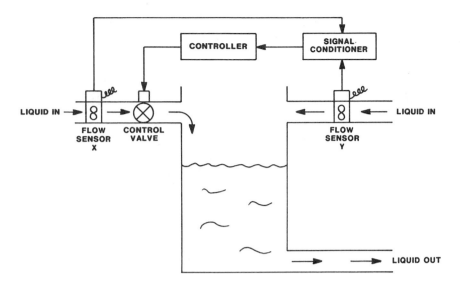

Figure 1–7   COMPOUND-VARIABLE PROCESS

are conditioned, then sent to the controller for evaluation. The two flow signals are scaled at the signal conditioner. Flow through port Y is left uncontrolled. Flow through port X is adjusted by the controller through the control valve at X to bring both input flows to a common ratio. As may be seen, there could have been two controllers, two control valves, and two sensors to perform this task. The comparison of flow information is a critical function in the process, as the result from the effort is to maintain the flow at some ratio, say 2:1. By leaving one flow unrestricted, the second may increase or decrease to satisfy the conditions of the ratio.

Cascade Process

The *cascade* process utilizes the measurement of one variable to establish the set point of a second variable. In this application the level of the liquid in the tank is determined by a liquid-level sensor. Level information is directed to the controller (see Figure 1–8). The controller has a set point of its own. The controller's output is the measurement of level. This output is used as level information to vary the set point of the flow control. Note that the flow control has its own sensor. Together, the level and flow information are integrated at the flow control to effect operation of the control valve. This action reflects the primary purpose of the application, which is to maintain a liquid level in the tank. The remarkable point about this application is that load changes are seldom, if ever, noticed because the flow control is in the process of regulating before the level becomes critical.

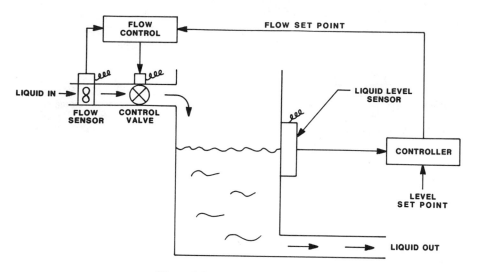

Figure 1–8   CASCADE PROCESS

2

Measurement

Measurement has to do with measuring or *mensuration*. Measurement is determining the extent, size, or quantity of something. Measurement is also a system of measuring or of measures. It is really not necessary to separate these three, for while working in the field of transducers we use them all collectively.

MEASUREMENT STANDARDS

Measuring both stimulus and response is common practice in testing the behavior of things. Today, at an accelerating pace, we are witnessing, experiencing, and intimately involved in the development and use of measuring instruments.

Transduction employs an energy transfer process to sense and communicate information. When a device transfers energy, involves a dual transaction, and manifests some physical law or effect, it is called a *transducer*. When a transducer functions primarily to sense and transfer information into a more convenient form, it is a *measuring transducer* or *sensor*. And, depending on its location in the structure of a system, a model can be an *input*, *modifying* (conditioning), or *output* transducer. In the measuring transaction, sensor structures obeying the laws of nature also tend to delay, distort, and degrade the initial information communicated and change the quantity being measured.

Since a measuring transaction involves a reaction and an energy transfer, it will always in some small way change the quantity being measured, and thereby affect the validity of the measurement. The purpose of the measurement—research, testing, control, or calibration—imposes different tolerances on this interaction effect. *Calibration* is testing the transfer behavior of a sensor in controlled transactions.

Measuring systems are composed of elements arranged in an organized structure and physically separated into component blocks. This model, patterned after human behavior, includes both the basic measuring process and related activities. The feedback path represents a basic mode of human behavior wherein the actual (measured) condition is compared with a desired (normal or predicted) condition and corrective, controlling, or adaptive action is taken according to the difference. For this process, one needs *norms and standards*.

International System of Units

The International System of Units (SI) has been established by international agreement. It has as its base six quantities: length, time, mass, temperature, electric current, and luminous intensity, with the corresponding units meter, kilogram, second, ampere, kelvin, and candela. All other SI units are derived from the six base units. A seventh base quantity, amount of substance (mol), is used in some fields.

Length.   The SI unit for length is the meter (m). The meter has been defined by agreement to be 1,650,763.73 wavelengths in vacuum of the orange-red line of the spectrum of krypton 86. One would use an interferometer to measure length by light waves. The meter has recently been redefined (1983) by international conference to be the length of a path traveled by light in a vacuum during a time interval of 1/299,792,458 part of a second.

Area is defined in terms of square meters (m^2). Volume is defined in terms of cubic meters (m^3).

Time.   The SI unit for time is the second (s). The second has been defined as the duration of 9,192,630,770 cycles of the radiation associated with a specific transition of the cesium atom. An oscillator is tuned to the resonant frequency of the cesium atoms as they pass through a system of magnets and a resonant cavity into a detector.

Frequency is defined in terms of cycles per second or hertz (Hz). Speed is defined by dividing distance (d) by time (t), d/t. The SI unit for speed is meters per second (m/s). Acceleration is defined as the rate of change in speed. The SI unit for acceleration is meters per second squared (m/s^2).

Mass.   The SI unit for mass is the kilogram (kg). The standard established is a cylinder of platinum–iridium alloy that is stored by the International Bureau of Weights and Measures near Paris, France.

Force is defined by the SI unit the newton (N). A force of 1 newton when applied for 1 second will give a 1 kilogram mass a speed of 1 meter per second. One newton equals approximately two-tenths of a pound of force.

The SI unit for work and energy is the joule (J).

$$J = N \times m$$

where

$$N = newtons$$

$$m = meters$$

The SI unit for power is the watt (W).

$$W = \frac{J}{s}$$

where

$$J = joule$$

$$s = second$$

Temperature.   The SI unit for temperature is the kelvin (K). The kelvin scale has a zero point fixed at absolute zero and has a point called a triple point at 273.16 K. The triple-point cell is an evacuated glass cylinder filled with pure water used to define a known fixed temperature. When the cell is cooled until a mantle of ice forms, the temperature at the interface of solid, liquid, and vapor is 0.01 degree Celsius (°C).

Fahrenheit (°F) is calculated as follows:

$$°F = 1.8(°C) + 32$$

Celsius is calculated as follows:

$$°C = \frac{°F - 32}{1.8}$$

Electric Current.   The SI unit for electrical current is the ampere. The ampere is the magnitude of current which when flowing through each of two parallel wires separated by 1 meter of space results in a magnetic force between the two wires of 2×10^{-7} newton for each meter of length.

Voltage or pressure is expressed in volts (V) and calculated as follows:

$$V = \frac{W}{A}$$

where

$$W = watts$$
$$A = amperes$$

Electrical resistance is expressed in ohms (Ω) and is calculated as follows:

$$\Omega = \frac{V}{A}$$

where

$$V = \text{volts}$$

$$A = \text{amperes}$$

Luminous Intensity.   The SI unit for luminous intensity is the candela (cd). The candela is the luminous intensity in a specific direction of a source that emits monochromatic radiation of frequency 540×10^{12} hertz and has a radiant intensity of 1/680 watt per steradian.

Light flux is measured in lumens (lm). A light source having an intensity of 1 candela in all directions radiates a light flux of 4π lumens.

Amount of Substance.   The SI unit for amount of substance is the mole (mol). The mole is the amount of substance that contains as many elementary entities as there are atoms in 0.012 kilogram of carbon 12.

Other Standards

Over the years, the United States and the world have developed standards that support the manufacturing and operation of equipment along with definition of specifications. These societies and professional organizations establish consistency in product and material. There are many; however, we identify only some of the best known in the field of transducers.

    ISA: Instrument Society of America
  ANSI: American National Standards Institute
SAMA: Scientific Apparatus Makers Association
  NBS: National Bureau of Standards
 BIPM: International Bureau of Weights and Measures
 IEEE: Institute of Electrical and Electronic Engineers
  IEC: International Electrotechnical Commission
  EIA: Electrical Industries Association
  ISO: International Standards Organization
   UL: Underwriters' Laboratory

Voltage standards.   Until the 1970s the unit of voltage standard was the *Weston standard cadmium cell*. The standard cell was accepted by the International Committee on Electrical Units Standards in 1908. In 1970, the unit of voltage became the *Josephson voltage standards*. These voltage standards are maintained by Josephson junctions. These are extremely low resistance junctions between two superconductors. This method may measure voltage of a group of standard cells with essentially zero-resistance parts. These standards were accepted and are maintained at the International Bureau of Weights and Measures (BIPM).

Resistance Standards. The resistance standard in the United States is the 1-ohm resistor designed by J. R. Thomas at the National Bureau of Standards (NBS). This resistor, which is hermetically sealed, wire-wound, and immersed in oil, is used by the NBS as a working standard.

Measurement Parameters

Table 2–1 lists the most common physical measurement parameters in use together with their symbols and SI units. The symbols provided are the preferred symbols used in most physics and engineering activities. The SI units are the Standard International Metric units. These are listed first, with the English or alternative units shown in parentheses. More common nomenclature is abbreviated. Table 2–2 lists the measurement parameter and the transduction principle popularly utilized.

TABLE 2–1 Physical Measurement Parameters

Measurement Parameter	Symbol	SI Unit (Alternative Unit)
Linear distance displacement, dimension, position	I, X, S	meter, m centimeter, cm millimeter, mm micrometer, μm (inch) (in.) (foot) (ft)
Linear velocity (speed)	$\dot{X}, V, \dot{S}$	meter per second, m/s kilometer per second, k/s (inch per second) (in./sec) (foot per second) (ft/sec)
Linear acceleration	$a, g, \ddot{X}$	meter per second per second (inch per second per second) (foot per second per second)
Angular displacement	θ	radian, rad (degrees) revolution (cycles) (arc-second)
Angular velocity (tachometer)	$\dot{\theta}$	radian per second, rad/s (revolution per minute) (rpm)
Angular acceleration	$\ddot{\theta}$	radian per second per second
Force, $F = ma$	F	newton, N kilogram-force dyne (pound foot) (lb-ft)

TABLE 2–1 (*continued*)

Measurement Parameter	Symbol	SI Unit (Alternative Unit)
Torque	T	newton-meter, N·m dyne-cm (pound foot) (lb-ft) (ounce inch) (oz-in.)
Vibration Displacement (distance, amplitude)	D peak to peak d peak	millimeter, mm (inch) (in.)
Velocity	V zero to peak	millimeter per second, mm/s (inch per second) (ips)
Acceleration	G, g, a	g meter per second per second (inch per second per second)
Pressure	Pa, p	pascal newton per square meter, N/m^2 (pound per square inch) (psi)
Flow meters	Q	cubic meter per second (cubic feet per minute) (cfm) gallon per minute (gpm) (cubic centimeter per minute) (cm^3/min) kilogram per second pound per minute pound per hour
Temperature	T	degree Celsius, °C (degree Fahrenheit) (°F) (degree Kelvin) (K) (degree Rankin) (°R)
Viscosity Dynamic Kinetic	Pa·s	Pascal second (poise) (P) (stoke) (ST)

TABLE 2–2 Measurement Parameters versus Transduction Principle Popularly Utilized

Measurement Parameter	Transduction Principle Popularly Utilized
Linear distance (displacement, dimension, position)	Capacitive, incremental or differential Inductive, eddy current or variable reluctance Strain gage, bonded or semiconductor

TABLE 2-2   (*continued*)

Measurement Parameter	Transduction Principle Popularly Utilized
	Linear voltage differential transformer (LVDT)
	Optical, photoelectric, or laser interferometer
	Linear potentiometer
	Linear (digital) encoders
Linear velocity (speed)	Inertial mass—magnetic field, self-generating
	Pendulous mass–spring
	Inductive
	Attached linkage
	Noncontact
	Optical—time differential
	Piezoelectric—integrated acceleration
Linear acceleration	Seismic mass
	Piezoelectric
	Piezoresistive
	Strain gage
	Inductive
	Capacitance
	Potentiometer
	Force balance servo
Angular displacement	Capacitive
	Inductive
	Potentiometric (resistance)
	Photoelectric
	Strain gage
	Rotary variable differential transformer (RVDT)
	Gyroscope
	Shaft encoder, digital
Angular velocity (tachometer)	Generator
	Dc
	Ac
	Drag cup
	Photoelectric or magnetic pulse wheel
Angular acceleration	Electromagnetic + differentiator
	Force balance servo
Force, $F = ma$	Counterbalance
	Mass
	Electromagnetic
	Deflection
	Strain gage
	LVDT
	Piezoresistive
	Capacitive
	Inductive
	Piezoelectric

TABLE 2–2   (*continued*)

Measurement Parameter	Transduction Principle Popularly Utilized
Torque	Torsional windup
	Strain gage
	Photoelectric encoder
	Permeability change
	Dynamometer
Vibration	Linear displacement transducer (DC-LVDT)
Displacement	Integrated linear velocity transducer signals
(distance, amplitude)	Double integrated accelerometer signals
Velocity	Linear velocity transducers, seismic
	Integrated accelerometer signals
Acceleration	Linear accelerometers, piezoelectric
Pressure	Bellows—potentiometers
	Capsule—differential transformer (LVDT)
	Diaphragm
	Strain gage
	Piezoresistive
	Piezoelectric
	Variable capacitance
	Variable inductance
Flow meters	Positive displacement, volumetric
	Liquid
	Gas
	Differential pressure
	Orifice
	Venturi
	Pitot tube
	Turbine—velocity
	Liquid
	Gas
	Magnetic—velocity
	Variable area
	Float meter
	Force meter
	Thermal—mass flow
	Differential pressure—mass flow
	Turbine-axial-momentum—mass flow
Temperature	Thermocouple
	K: Chromel–alumel
	J: Iron–constantan
	B: Platinum–rhodium
	T: Copper–constantan
	RTD
	Platinum
	Nickel
	Thermistor
	Semiconductor junction

TABLE 2–2 (*continued*)

Measurement Parameter	Transduction Principle Popularly Utilized
	Pyrometer
	Radiation
	Optical
Viscosity	Falling
Dynamic (poise)	Ball
Kinetic (stoke)	Piston
	Capillary or orifice (Saybolt)
	Rotating member

MEASUREMENT AND RANGE DEFINED

Measured Variable

The physical quantity, extent, or size of the entity being measured is often called the measurand. An example of the measured variable is the pressure in an aircraft hydraulic actuator. This pressure could range from 0 to 3000 psi. That pressure is the measured variable.

Measured Signal

The measured signal is the analog of the measured variable produced by the transducer. For example, the aircraft hydraulic actuator has a high measured variable of 3000 psi. The measured signal produced by the transducer may be 5 V at the high range, representing 3000 psi. The measured signal may also be a combination of signals representative of the total of the measured variable. Two or three fuel flows could be combined and measured as one measured signal.

Range

A transducer is usually constructed to operate within a specific range. A thermocouple could not, for example, be expected to operate with efficiency in a range from 0 to 1,000,000°C. It could, however, operate extremely efficiently in a short range of, say, 20 to 140°C. Any temperature under 20°C would be *under range*. Any temperature over 140°C would be *over range*. Midrange would be at 80°C. In most cases, transducers are manufactured so that they are linear at certain levels of the measured variable. Curves are provided to the buyer so that they know where and when the transducer operates in the most linear manner. Range, then, is the region between the limits wherein a variable is measured, received, or transmitted.

Span

Span is related to range. Span is the algebraic difference between lower and upper range limits. For example, in the range of two temperatures, -25 to $100°F$:

$-25°F$ is the lower limit;

$100°F$ is the upper limit;

$125°F$ is the span.

Zero Point

The zero point is important when collecting data. It is the starting point at which a variable is to be measured. For example, a pressure gage could be zeroed at atmospheric pressure.

STATIC ACCURACY DEFINED

Accuracy

The accuracy of a transducer is the maximum error that the user may expect from a transducer. That accuracy is all-inclusive of the various errors. For example, a certain amount of error may be witnessed when measuring each temperature. Another error may be expected at upper or lower scale. Still another temperature error may be expected at different pressure ranges. All errors must be combined to estimate or calculate the total accuracy of a device or system. Usually, the manufacturer will provide an accuracy rating in percent, such as 0.02% over the span of measurement or a $\pm$ rating, such as $\pm3°F$.

Error

Error is the algebraic difference between the actual value of a variable and the set point or indicated value. That error is provided by the transducer manufacturer. If you subtract the error rating from the value indicated, you will obtain the actual value:

actual value = indicated value − error

Error may be negative or positive. Error is often expressed in percent relative to the set point. For example, if the pressure in a system were set at 2800 psi and the measured pressure was 3000 psi, the error would be 200 psi. As a percentage it would appear thus:

$$\frac{\text{error}}{\text{set point}} \times 100\% = \% \text{ of error (of set point)}$$

$$\frac{200 \text{ psi}}{2800 \text{ psi}} \times 100\% = 7.14\%$$

Error is also expressed as a percentage relative to the range. For example, if the pressure range of a pressurized tank were from 1000 to 3000 psi, the span of the range would be 2000 psi. If there were an error in set point of 200 psi, the percent of error would be 10%.

$$\frac{\text{error}}{\text{range max.} - \text{range min.}} \times 100 = \% \text{ of error (of span)}$$

$$\frac{200 \text{ psi}}{2000 \text{ psi}} \times 100\% = 10\%$$

In the two cases above, the percent of error looks larger for the latter, but interpretation would provide obvious conclusions.

Linearity

The most desirable function a transducer may have is to be linear throughout its span of operation. That is, it must have an output that is directly proportional to its input throughout its entire range of operation. Figure 2–1 is a linearity curve. The solid line represents the expected (desired) results from a transducer. The dashed line is the actual value produced at various inputs. *Deviation* is the difference between the two curves. Obviously, the desired output is a straight-line function. Since this is rarely achieved, the best case would be some value within a maximum deviation. Straight-line response is linear. Variable or bent-line response is nonlinear.

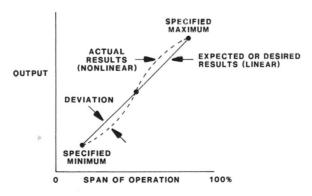

Figure 2–1 LINEARITY

Hysteresis

Hysteresis is recognized when a transducer's output is monitored in a cyclical manner. It becomes apparent that the output with increasing input is different from the output with decreasing input. In Figure 2–2 the decreasing curve is nearer the straight line than is the increasing curve. Further cycling should repeat the same curves. The two parts of the curve will be one loop called a *hysteresis loop*. Hysteresis is the total deviation from X to X.

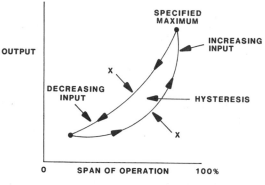

Figure 2–2   HYSTERESIS

Deadband

Often in a hysteresis loop there is an area within the input that does not vary the output by an amount that is observable. That area is called the deadband. The transducer may operate in this range without expecting any change in output response for a given input. The deadband in Figure 2–3 is exaggerated for clarity.

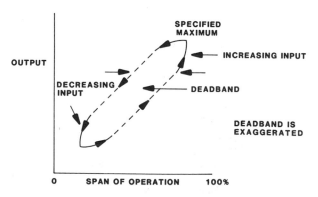

Figure 2–3   DEADBAND

Sensitivity

Sensitivity is the amount of change in output of a sensor as a result of a change in input. High sensitivity is desirable. High sensitivity means that a large output may be had with a small-signal input. Low sensitivity means that a low output would result from a small-signal input. For example, the sensitivity of a pressure transducer may be 1 millivolt per 100 psi (1 mV/100 psi). Sensitivity may also be specified in terms of excitation voltage against full-range value, that is (as an example), 3 millivolts per volt (3 mV/V) excitation at full pressure.

Repeatability

Repeatability represents the ability of the transducer to repeat itself each time an input is applied. For every input an expected output is achieved.

Resolution

The resolution of a transducer is the smallest change that can be detected by the transducer.

Deviation

Any expected or nonexpected change from the normal operation, a linear operation, or the desired operation of a transducer is called deviation.

Conformity

Conformity is representative of how much or how little the outputs of a transducer appear to be similar to specified values.

Drift

If a signal level of a transducer varies from its zero point, that variation is called drift. Drift is caused by aging of components or by some physical phenomena. It may only appear for a short time, thereby being called *short-term drift*. *Long-term drift* would last for an extended period.

DYNAMIC ACCURACY DEFINED

Set Point

The set point (see Figure 2–4) is the dynamic variable level at which a process is to operate. The set point is selected by the process engineer to represent that value of dynamic variable which best expresses a desirable parameter. This

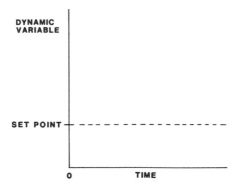

Figure 2–4   SET POINT

value may be 100°F when used with a thermocouple, 3000 psi in use with a pressure transducer, $\frac{3}{8}$ in. using a displacement transducer, and so on. Whatever the use may be, the control set point (expressed as C_{SP}) is preselected as the value to work at, to, and from. A measurement of some physical phenomenon is converted to a current level, then compared with the set point of a current expressed in the same proportion. It is from this evaluation of measured versus set point that the control system begins its work of control.

Dynamic Control

Regulation of a control system and its process variable should be smooth, with only minor variations from the set point (see Figure 2–5). If the loop is operating in the correct manner, any deviation about the set point will be small, with only minor excursions. The better the loop, the less the error. *Dynamic response* is that measure of a system operation as a function of time. Response is automatic and intentional to keep the loop operating about the set point.

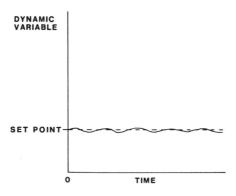

Figure 2–5   DYNAMIC CONTROL

Transients

A transient is an unexpected signal that appears at the control loop. The problem with this condition is that the control loop looks at the transient as another change in the controlled variable. The control loop must attempt to correct for the change. The attempt is called *transient response*. Transient response is corrected by either underdamping or overdamping. *Underdamping* causes the signal to oscillate above and below the set point with decreasing amplitude until the set point is reached (see Figure 2–6). This mode is often called the *cyclic method*. The cyclic method may be used when settling to the set point is not critical. *Overdamping* is utilized when time is not critical, but it is imperative that no oscillations occur beyond the set point (see Figure 2–7). This mode is called the *damped* mode.

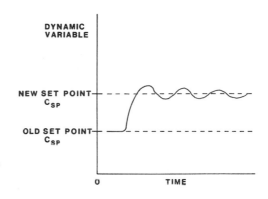

Figure 2–6 CYCLIC CHANGE IN SET POINT

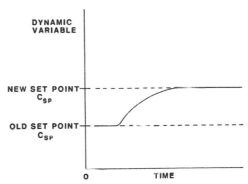

Figure 2–7 DAMPED CHANGE IN SET POINT

Time Response

The underlying fact of transducer use is that all the variable information that is measured or monitored must be converted to an analog form that is acceptable for conversion to digital form. That analog form is often time dependent. If the control system must be time regulated, the engineer must design around time lags, even for gradual or step responses. There are two types of time response, first- and second-order time lag. *First-order response* is distinguished by the fact that the control loop simply cannot respond to rapid or step changes in inputs. Therefore, the response of the output will gradually lag until the set point is reached. *Second-order response* (or lag) is described by oscillation about the set point. Oscillation is highly irregular, and error information will exist until the oscillation or second-order time response decays to the set point.

Settling Time

Settling time is the time required for the variable to settle to the new set point. This time allows for the accepted deviation at the set point.

Rise Time

Rise time is the time required for the variable to move from a small level near the old set point to a higher level near the new set point. These points are specified as percentages, such as between 5 and 95%.

Residual Error

Residual error is seen after the move from old to new set points (see Figure 2–8). It is the amount of difference between the desirable set point level and the point at which the dynamic variable has stabilized.

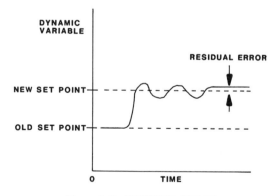

Figure 2–8   RESIDUAL ERROR

Response Time

If an output is expressed as a function of time and its input is specified under specific operating conditions, the result is called response time.

Noise

Unwanted signals that are part of the variable or are being carried as part of the variable are called noise. The noise is usually detrimental to the reception of signal information.

Frequency Response

Frequency response is a complex method for analyzing the stability of a control loop. The operation involves insertion of sinusoidal input signals into the control loop and monitoring the gain and phase shift of its output. Input and output are compared through a large range of frequencies to determine the changes of gain and phase shift under various conditions. Response curves are created for components or total systems. This technique for analyzing stability is a complex and often costly procedure. Test engineers must develop setups that utilize expensive equipment and require much precious time. If, however, the control loop is on a system such as an aircraft flight control, it may save fuel or even lives.

Stability

In all cases, the dynamic variable is not permitted to increase beyond limits without some sort of control. Furthermore, the stability of a control system demands that the variable not oscillate at an uncontrolled amplitude. A system is stable if it can control its gain and frequency response. These are major subjects and must be studied in depth to be comprehensible. That depth is beyond the form of simple definition and far beyond the scope of this book.

DYNAMICS OF NEGATIVE FEEDBACK

In general terms, feedback is a part of the output signal that is fed back into an amplifier to achieve stability, remove distortion, and cancel noise in the closed loop of an amplifier. In closed-loop control systems, negative (degenerative) feedback is used purposefully to reduce the effect of internal and external disturbances.

The effect of feedback on the control system circuit is a loss of gain and alteration of the input impedance. Since the amplifiers used in today's electronics have massive gain and input impedances in the millions, these negative effects are really not worthy of concern. The positive effect of feedback is the improved performance of the control system. Negative feedback opposes a change in the input signal and tends to maintain that input at a level requested by the set point reference.

Feedback in a basic control system may be represented by the feedback concept illustrated in Figure 2–9. In this figure the dynamic variable is V_{IN}, with the output variable labeled V_{OUT}. An amplifier with the gain of A is in series

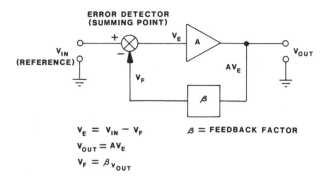

Figure 2–9   FEEDBACK CONCEPT

with an error detector (summing point). The feedback factor (β) is connected between the output and the input summing point. A change in input variable takes place. That change is summed with the negative feedback voltage V_F, then applied. The amplified voltage provides an output voltage V_{OUT}. The output voltage is fed through the feedback factor (β) component(s), where it becomes feedback voltage (V_F). V_F is summed with the input variable V_{IN} at the error detector (summing point).

V_E, the input to the amplifier, is a function of input voltage and feedback voltage. V_E is the difference between these two voltages.

$$V_E = V_{IN} - V_F$$

V_{OUT} is a function of the input to the amplifier and the gain of the amplifier.

$$V_{OUT} = AV_E$$

V_{OUT} may also be illustrated as a function of the amplifier gain and the difference (error) between the input voltage and the feedback voltage.

$$V_{OUT} = A(V_{IN} - \beta V_{OUT})$$

or

$$V_{OUT} = AV_{IN} - A\beta V_{OUT}$$

Also,

$$V_{OUT} + A\beta V_{OUT} = AV_{IN}$$

and

$$V_{OUT}(1 + A\beta) = AV_{IN}$$

Therefore,

$$\frac{V_{OUT}}{V_{IN}} = \frac{A}{1 + A\beta}$$

Since feedback is applied and V_{OUT} divided by V_{IN} is the gain of the control system, feedback gain is

$$\frac{V_{OUT}}{V_{IN}} = \frac{A}{1 + A\beta} = A_{FB}$$

where A_{FB} is the closed-loop gain of the amplifier with feedback. The feedback factor β is a portion or factor of the output fed back to the input. The actual component is usually a resistor but may be a capacitor or inductor.

3

Transducers

This chapter describes the transducer and the elements that allow transducers to accomplish their task. The importance of this chapter is that the reader realize the function that must be dealt with. By understanding the elements, the task of interfacing becomes simpler. In addition, the knowledge of how a transducer functions may aid in mechanizing it to the force to be measured and support the transducer's isolation from other environmental inputs.

TRANSDUCER TYPES

One of the functions of a transducer is to detect and convert a measurand into an electrical quantity. The measurand is the material or energy being measured, such as the amount of coal in a hopper or the pressure within a closed hydraulic system. It is also worth noting that the measurand need not be energy, but could be material quantities. The other function of a transducer is the control function. This involves a system in which the transducer is an integral part, as implied in Chapter 1.

There are basically two wide areas into which transducers are categorized: active and passive. *Active transducers* are those that generate a voltage or current as a result of some form of energy or force change. For example, a thermocouple, when heated, generates an electrical signal that is proportional to the amount of heat applied. The *passive transducer* changes its properties when exposed to energy. For example, photoresistors change resistance when exposed to light energy. Therefore, the current through the photoresistor changes. In this event we can again say that the amount of electrical output change is dependent on the amount of energy (light) to which the photoresistor was exposed.

There are other special transducer elements that may embody the active or passive elements for some combination of special interest. The electrokinetic transducer, for example, operates with polar fluid maintained between two diaphragms to change the difference of potential between faces of its electrical interface. This element does not quite fit into the categories of active or passive transducers.

Thus although transducers usually fall into the active or passive categories, some employ special combinations of characteristics that set them aside from the standards generally denoting active or passive. Such transducers include the electrokinetic, force balance, oscillator, differential transformer, photoelectric, vibrating wire, velocity, and strain gage element.

ACTIVE TRANSDUCER ELEMENTS

Active transducer elements are usually graded as those elements that generate a voltage as a result of some energy or force change. The generation of signals is accomplished by six major methods. Two of these, *electrostatics* and *chemical*, are not in general use as transducer element types. The other four methods are the *electromechanical*, *photoelectrical*, *piezoelectrical*, and *thermoelectrical*.

Electromechanical Transducer Elements

Relative motion through a magnetic field will produce a voltage at the ends of the conductor. That is, if a conductor passes through a field or the field is moved across a conductor, the motion will produce a voltage at the ends of the conductor. Such is still the case if both the conductor and the magnetic field are moving.

All these statements are representative of *Faraday's law of induction*. In the 1880s, Michael Faraday, an English scientist, developed a simple machine in which a conductive disk was rotated in a magnetic field. Sliding contacts (brushes) were used to pick off the voltage from the disk. The mode of operation is called a single-pole generator. In Figure 3–1A a fixed-level conductor is moved through a field. The motion produces a constant level of voltage across the conductor. In Figure 3–1B the conductor is rotated in the magnetic field, thereby producing an alternating voltage. The magnitude of the voltage generated (E) is dependent on the flux density (B) of the magnet, the length of the conductor (L), and the speed of the conductor (V):

$$E = BLV$$

Faraday's rotational conductor was somewhat different. An equation was developed as a result of Faraday's law of induction. The induced voltage (V)

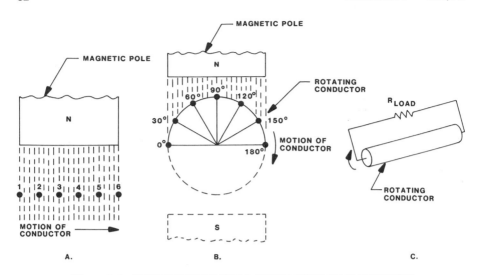

Figure 3–1   ELECTROMECHANICAL GENERATION OF ELECTRICITY

produced in a conductor increases as the number of conductors (N) moving through the field increases:

$$V \propto N \frac{d\phi}{dt}$$

where

V = induced voltage

$\propto$ = proportional to

N = number of conductors

$d\theta/dt$ = rate at which the magnetic flux crosses the conductor

The equation suggests that induced voltage increases whenever the number of poles increases or the rate of motion of the conductor through the magnetic field increases.

The reader will note that Faraday's law of induced voltage changes somewhat with a loaded circuit. That is, when a load resistance is added to the conductor as in Figure 3–1C, a current is induced through the conductor and load which opposes the motion of the conductor in the magnetic field. Without the load resistance, the conductor moves easily through the field. With the load resistance installed, the conductor does not move easily. The reason for this is that the current through the conductor produces a magnetic field which opposes the motion that generated the voltage in the first place. *Lenz's law* modifies the Faraday equation and can be stated mathematically as

$$V = -N \frac{d\theta}{dt}$$

Note the similarity to the induced voltage equation stated previously. The minus (−) sign denotes the opposition developed by the load resistance.

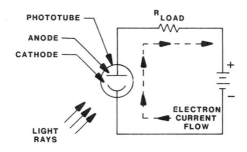

NOTE: CATHODE IS COATED WITH RADIATION
SENSITIVE METAL. ELECTRONS ARE
RELEASED AS RADIATION IS APPLIED.

Figure 3–2   PHOTOEMISSION

Photoelectrical Transducer Element

The effect of light on conductive material produces an effect called the *photoelectric effect*. The sensing of light is also called *photodetection*. Photodetection is defined around three phenomena: *photoemission*, *photoconduction*, and *photovoltaic actions*. All quantum photodetectors respond directly to the action of incident light. The first, photoemission, involves incident light, which frees electrons from a detector's surface (see Figure 3–2). This usually occurs in a vacuum tube. Note that electron current flows from negative to positive. With photoconduction, the incident light on a photosensitive material causes the material to alter its conduction (see Figure 3–3). The third phenomenon, photovoltaic action, generates a voltage when light strikes the sensitive material of the photodetector (see Figure 3–4).

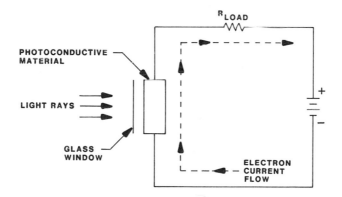

NOTE: PHOTOCONDUCTIVE MATERIAL DECREASES
RESISTANCE AS RADIATION IS APPLIED.

Figure 3–3   PHOTOCONDUCTION

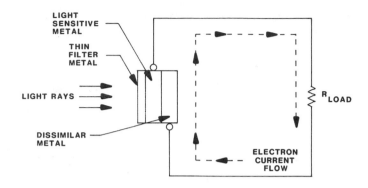

LIGHT
SENSITIVE
METAL

THIN
FILTER
METAL

LIGHT RAYS

DISSIMILAR
METAL

R LOAD

ELECTRON
CURRENT
FLOW

NOTE: RADIATION CAUSES DIFFERENCE
OF POTENTIAL BETWEEN TERMINALS.

Figure 3–4 PHOTOVOLTAIC ACTION

Note that there is no power source such as a battery involved with photovoltaic action.

Piezoelectrical Transducer Elements

Synthetic or natural crystals such as quartz have special physical qualities. When crystalline material is stressed (pressure causes the crystal shape to change or distort), a voltage is produced at the surface. If a material performs in this manner, it is said to be *piezoelectric*. Piezoelectric material also has another quality that is the direct reverse of piezoelectric. This second quality is *electrostriction*. Electrostriction involves the application of an electrical field to the crystal substance. The electrical field alters the crystal shape. Probably the most common example of piezoelectric material is the ceramic material in the stylus of a record player. As a record turns, the ceramic crystal flexes in the grooves, causing a voltage that represents the amount of flex. The voltage is then amplified and sent to a loudspeaker. Another common use of the piezoelectric effect is the control of radio frequencies in transmitters.

A most basic piezoelectric transducer is illustrated in Figure 3–5. In the figure, quartz with a beam is tied to the support base. The crystal and its beam are connected mechanically to a pressure-sensing diaphragm which acts as a force summing device. The diaphragm, in turn, is open to a pressure port. A change in pressure causes a mechanical change on the crystal's beam, which causes the crystal to oscillate at a specific frequency or to generate an electrostatic charge signal. This is, of course, extremely fundamental.

These devices have high-frequency response, are self-generating, and are small and rugged. They are however, sensitive to temperature changes and cross-accelerations, have high-impedance outputs, and do not recover very quickly after extreme shock.

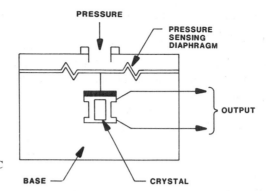

Figure 3–5 PIEZOELECTRIC
TRANSDUCER ELEMENT

Thermoelectric Transducer Element

The sensing of temperature is usually accomplished with the aid of a thermo-electric device such as the thermocouple, the resistance temperature detector (RTD), or the thermistor. Whichever device is used, it plays a major role in the monitoring of heat energy, such as in conversions from coal to steam and air conditioning.

There are three thermoelectric effects that play a large part in generation of electricity by way of temperature. The first is the *Seebeck effect* (see Figure 3–6A). Thomas Seebeck, a German physicist, fused two dissimilar metal wires together on both of their ends. He then heated one of the junctions and found that electrical current flowed from one wire to the other. He caused electrons to flow from a

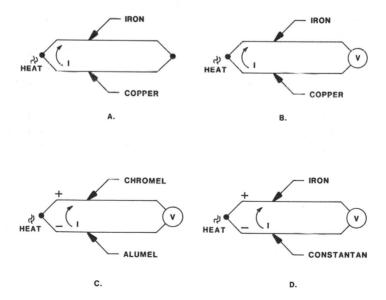

Figure 3–6 (A) SEEBECK EFFECT (B) MEASURING THERMOCOUPLE
POTENTIAL (C) (D) ELECTRICAL POLARITY

copper wire to an iron wire. This effect developed into what we know now as the *thermocouple*.

The heated junction is called the hot junction; the other is called the cold junction. In Figure 3–6B the cold junction is replaced with a voltmeter. The voltmeter provides a closed-circuit path and monitors the difference of potential across the heated junction.

The potential developed across the heated junction is the thermocouple potential. Its polarity and magnitude are dependent on the type of material in the two dissimilar metals (see Figure 3–6C and D). In Figure 3–6C the positive polarity is on the chromel side of the thermocouple and the negative polarity is on the alumel side. In a second example in Figure 3–6D, the positive polarity is on the iron while the negative polarity is on the constantan. The metal chromel is an alloy of nickel and chromium. Alumel is an alloy of nickel, magnesium, aluminum, and silicon. Constantan is an alloy of copper and nickel.

The second thermoelectric effect is called the *Peltier effect*. Jean Peltier, a French physicist, applied current to a junction made by two dissimilar materials (see Figure 3–7A). In the illustration, an iron and a copper wire are fused together at their ends. A battery is placed in series with the iron lead. Current flows in the entire closed-loop circuit. As electrons flow from the iron material to the copper material, the junction (A) becomes hot. This is because the electrons are moving from a high-energy-state material, iron, to a low-energy-state material, copper. The excess energy heats the junction. As electrons flow from the copper material to the iron material, the junction (B) becomes cold. This is because the electrons are moving from a low-energy-state material (copper) to a high-energy-state material (iron). The thermal energy of the junction supplies the energy for transition. If the current were reversed, as in Figure 3–7B, the junction (B) would become the hot junction, while junction (A) would become the cold junction. The phenomenon would not change, however. The hot junction will always be where electrons are moving from a high-energy-state material to a lower-energy-state material.

Although the Peltier effect in application does not constitute a transducer or a sensor, it is appropriate to mention it as a *thermoelectric effect*. Its applications are with devices that require cooling or heating functions. The Peltier effect is

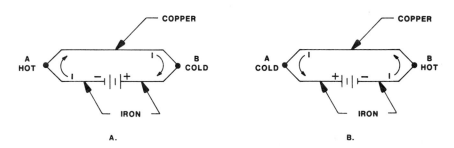

Figure 3–7 PELTIER EFFECT

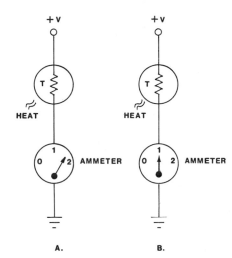

Figure 3–8 FARADAY EFFECT: (A) HEAT APPLIED (RESISTANCE DECREASE, CURRENT INCREASE) (B) HEAT REMOVED (RESISTANCE INCREASE, CURRENT DECREASE)

applied with semiconductor materials such as bismuth telluride as a conductor of thermoelectric carriers.

The third thermoelectric effect is called the *Faraday effect*. Faraday found through experimentation that certain semiconductor materials decrease their resistance as temperature increases. The material is said to have a *negative temperature coefficient*. It was found later that oxides of cobalt, manganese, and nickel provide thermally sensitive resistance for temperature-involved applications. The resistance became known as the *thermistor*. In Figure 3–8A a thermistor is installed in a circuit with an ammeter. Heat is applied to the thermistor. The thermistor decreases resistance. Current flow increases and the ammeter reflects the increase. When heat is taken away, the resistance increases and current decreases (see Figure 3–8B). The thermistor has, then, sensed the change in heat so that it can be monitored as current flow.

PASSIVE TRANSDUCER ELEMENTS

Each *passive transducer* has an element which, under some force, responds by moving mechanically to cause an electrical change. Three of these are the *capacitive*, *inductive*, and *resistive* transducer elements.

Capacitive Transducer Elements

The capacitor consists of a pair of plates made of conductive material placed on each side of or around a nonconducting material (insulator) known as a *dielectric* (see Figure 3–9A). Leads from each conductive plate are connected to the circuit. The capacitor element operates in the same manner as does a simple capacitor (see Figure 3–9B). Force from a measurand (that which is being measured) moves

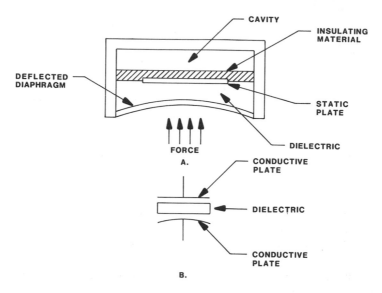

Figure 3-9   CAPACITIVE TRANSDUCER

one or both of the plates, which changes the capacitance of the capacitor. In an
excited circuit, the capacitance change would cause a change in capacitive reactance,
which, in turn, would modify current flow.

Consider what may happen in the event of a measurand change. The exact
formula for calculating the capacitance of a capacitor depends on the capacitor's
size, shape, and *dielectric constant*. The basic formula for calculating the capacitance
of a parallel-plate capacitor is

$$C = \frac{kA}{D}$$

where

C = capacitance, farads

k = dielectric constant

A = area of either plate

D = distance between the plates

The formula may be expanded in the following manner:

$$C = \frac{kA}{D} \times 22.4 \times 10^{-14}$$

where the value 22.4×10^{-14} farad is used to convert the C to a farad unit. The
formula indicates that any of the physical factors that are involved in the capacitor's
makeup could be used to change its capacitance.

The dielectric constant is dependent on the material from which the dielectric is made. Typical of the dielectrics are vacuum, paper, mica, ceramics, and glass. Vacuum is the base dielectric constant (1); other examples are air (1.0006), paper (2), mica (3), glass (8), and ceramics (100).

Another equation that is important to the capacitive transducer is the quantity of charge (Q) of electricity that can be stored in the capacitor. The formula is

$$Q = CV$$

where

Q = quantity of charge, coulombs

C = capacitance of the capacitor

V = voltage applied, volts

The charge on a capacitor is made by repelling free electrons from one plate of the capacitor ($+$) and attracting them to the opposing plate ($-$). This causes a difference of potential across the dielectric equal to the applied voltage. To discharge a capacitor, a short is placed across the dielectric. The action allows free electrons to move back to atoms that have open holes in their valence rings, thereby stabilizing the atoms.

The time to charge a capacitor is important in capacitive circuits and is used as a parameter in transducer applications. In general, the capacitor is used along with circuit resistance to determine the *time constant* of the circuit. The time constant in a resistive–capacitive (*RC*) circuit is the time in seconds that it takes for the capacitor in the circuit to charge to 63.2%.

An *RC* circuit contains at least one capacitor and one resistor. Figure 3–10A illustrates this circuit with a voltage input of 5 V. When the switch is closed, current begins to charge the capacitor. The capacitor charges by 63.2% of the applied voltage, as follows:

Applied voltage	5.00 V
First time constant	− 3.16 V
Remaining voltage	1.84 V

During the second time constant, voltage will increase 63.2% of 1.84 V: 0.632 × 1.84 = 1.16 V increase.

The charge voltage on third, fourth, and fifth time constants increases in the same manner. After the fifth time constant, for practical application, the charge has reached the full 5.0 V applied. While the charge is building up from 0 to 5 V, the capacitor is opposing a change in voltage. The reaction of the *RC* circuit during this time is called *transient response*. The time constant (TC) in seconds is a ratio of capacitance and resistance. Figure 3–10B and C are curves that represent charge and discharge time, respectively:

$$TC = RC$$

$$= 5000 \times 0.0001 \ (5 \ k\Omega \times 100 \ \mu F)$$

If the circuit shown in Figure 3–10A is opened, the reaction of the circuit will be that the capacitor will discharge through the resistor by 63.2% during the first time constant, then by 63.2% of the remaining charge during the second time constant. The capacitor will continue to discharge in the same manner, and after the fifth time constant will, practically speaking, have a zero charge. The time constant grows in value if either the capacitance or the resistance is increased. Conversely, the time constant becomes less if either the capacitance or the resistance is decreased.

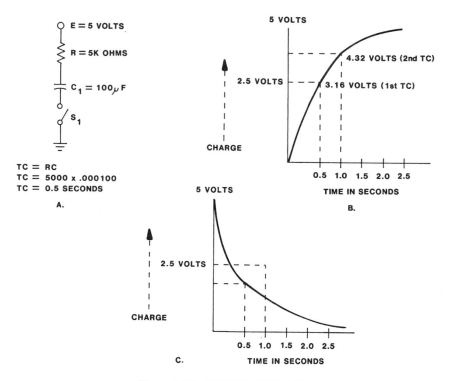

Figure 3–10 *RC* TIME CONSTANT

Inductive Transducer Elements

The inductive transducer element consists of a diaphragm or core called an *armature* that is driven by force from a measurand. The armature in the left view of Figure 3–11A is either displaced or rotated near the coils of a C-shaped pickoff. The displacement or rotation of the armature causes a change of inductance of the magnetic flux in the coil by varying the air gap in the flux path. In an

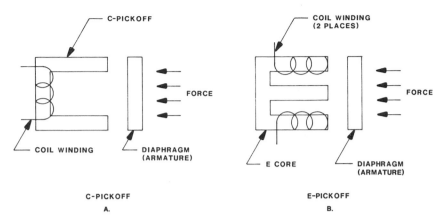

Figure 3–11 INDUCTIVE TRANSDUCER ELEMENT: (A) C-PICKOFF
(B) E-PICKOFF

excited circuit this would cause a change in inductive reactance, which, in turn, would modify current flow.

In Figure 3–11B the armature is either displaced or rotated between a pair of coils of an E-pickoff. The displacement or rotation of the armature causes a change of inductance between the two coils.

Let's consider what may happen in the event of a measurand change. The parameters that are involved in the physical makeup of the coil can be modified by the force of the measurand. The inductance of a coil or inductor may be changed by the core material, the relationship between the armature and the E-pickoff, the number of coil turns, the diameter of coil, and the coil length.

An inductor operates on the principle of induction. Current through an inductor lags voltage by 90° in a pure inductive circuit and by some angle between 0 and 90° in other *RL* circuits. As current increases through an inductor, a magnetic field is built up around the coil. The strength of this field is dependent on the physical makeup of the inductor and the core permeability. The magnetic field produces a counterelectromotive force which opposes a change in current.

The core of the inductive pickup has some properties that cause some error. The most prominent of these losses are in *eddy currents* and *hysteresis*. Eddy currents are prevalent in iron cores. Alternating current induces voltage in the core, causing currents to flow in a circular path through the core. This causes a loss in signal power. Eddy currents occur more often at high-frequency ac operation. The second type of loss is hysteresis. Hysteresis also occurs most prominently at high frequencies; it is loss of power as a result of switching or reversing the magnetic field in magnetic materials.

The *Q* of a coil represents its ability to store internal energy. The factor is a ratio between the inductive reactance of the coil and its internal resistance:

$$Q = \frac{X_L}{R_L}$$

where

Q = quality or merit of the coil

X_L = inductive reactance of the coil R_L

R_L = internal resistance of the coil

At low frequencies, the internal resistance of the coil is purely the dc resistance of the wire. At high frequencies, power losses from eddy currents increase and the Q of the coil decreases. Q of a coil can be identified by the ratio of reactive power to resistive or real power:

$$Q = \frac{P_{XL}}{P_R}$$

where

Q = quality or merit of the coil

P_{XL} = reactive power

P_R = resistive power

Significantly, the time that it takes for the field to build up and collapse in an inductive circuit is an important factor in transducer applications. In general, the inductor may be used as a timing component along with the resistive elements in a circuit.

The *time constant* in a resistive–inductive circuit is actually the time that it takes for the current to change by 63.2%. An *RL* circuit contains at least one inductor and one resistor. Figure 3–12A illustrates this circuit, which has 1.0 A of steady-state current flow.

When the switch is closed, current starts to flow and in one time constant increases from 0 to 0.632 A. (See Figure 3–12B for a curve that represents the field being built up and the relationship between curve and time.) During the second time constant, current increases by 63.2% of the remaining steady-state current, as follows:

Steady-state current	1.000 A
First time constant	− 0.632 A
Remaining	0.368 A

63.2% of 0.368 = 0.232 A increase during the second time constant

Current continues to increase for the third, fourth, and fifth time constants in the same manner. After the fifth time constant, for practical application, current has reached the 1.0-A steady-state current.

While current is building up from 0 to 1.0 A, the inductor is opposing the

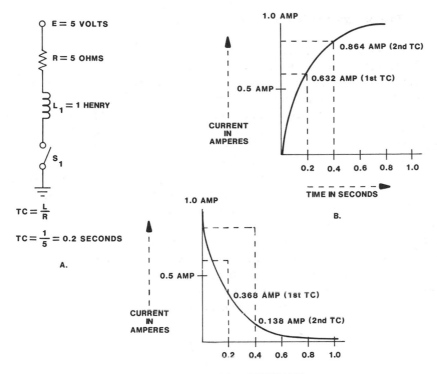

Figure 3–12 *RL* TIME CONSTANT

change in current. The reaction of the *RL* circuit during this time is called *transient response*. The time constant (TC) in seconds is a ratio of inductance and resistance, as follows:

$$TC = \frac{L}{R}$$

$$= \frac{1 \text{ H}}{5 \text{ }\Omega}$$

$$= 0.2 \text{ s}$$

If the circuit shown in Figure 3–12A is now opened, the reaction of the circuit will be that the current will decay 63.2% in the first time constant, then by 63.2% of the remaining current in the second time constant.

The current will continue to decay in the same manner, and after the fifth time constant will, practically speaking, be zero. (See Figure 3–12C for a curve that represents the field collapsing and the relationship between current and time.) The time constant grows larger with an increase in inductance and smaller with an increase in resistance.

Potentiometric Transducer Elements

Resistive transducers may be potentiometric. There are other resistive elements available, such as the slide resistor or other forms of variable resistance. A *potentiometric transducer* is an electromechanical device consisting of a resistive element with a movable wiper or slider (see Figure 3–13). A measurand causes a force to act on the wiper. The wiper makes contact along the resistance in relation to the amount of force applied by the measurand. As the wiper moves, the output may be taken from between one end of the resistance and the wiper. Output values may be linear, trigonometric, logarithmic, or exponential.

The potentiometric transducer is usually large in size, although there has been considerable progress in miniaturization. The potentiometric transducer also has high friction problems and is extremely sensitive to vibration. It develops high noise as it ages and has low-frequency response. Despite these problems, the potentiometric transducer is inexpensive compared to other transducers. The device is easily installed and can be excited by both alternating current (ac) and direct current (dc). Finally, the potentiometric transducer has a high output with no amplification or impedance-matching problems.

The basic formula involved with the potentiometric transducer input and output voltages is as follows:

$$V_{\text{out}} = V_{\text{in}} \frac{R_2}{R_1 + R_2}$$

This is a voltage-divider formula whose output is taken from R_2, the lower-side resistance as picked off by the wiper (see Figure 3–13).

If the output were taken from resistance R_1 (the upper side of the wiper),

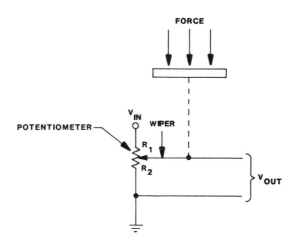

Figure 3–13 POTENTIOMETRIC TRANSDUCER ELEMENT

the polarity of the output would change ends and the voltage-divider formula would change slightly:

$$V_{\text{out}} = V_{\text{in}} \frac{R_1}{R_1 + R_2}$$

SOME SPECIAL TRANSDUCER ELEMENTS

Some transducers employ the basic action of active or passive transducers. Others may employ a special combination of characteristics that set it partly aside from the standards determining active or passive qualities. We will examine some of these transducers.

Electrokinetic Transducer Elements

An electrokinetic transducer consists of polar fluid contained between a pair of diaphragms (see Figure 3–14). A porous disk is inserted in the polar fluid between two partially porous plates called *diaphragms*. A change in force from the measurand causes the diaphragm to deflect and allow a very small amount of polar fluid to flow through the porous disk partition to effect a deflection on the second diaphragm (on the left in the illustration). This flow causes a difference of potential between the plates of the porous plug. A second effect may be recognized in reverse operation. An electrical potential may be applied to plates of the porous plug, which in turn causes polar fluid flow and diaphragm deflection.

The electrokinetic transducer element is self-generating, with relatively high frequency response and high output. It does have disadvantages, in that it cannot monitor static pressures or linear accelerations. The element cannot be calibrated without flow. A further disadvantage is that the polar fluid is often volatile.

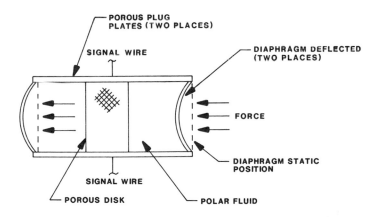

Figure 3–14 ELECTROKINETIC TRANSDUCER ELEMENT

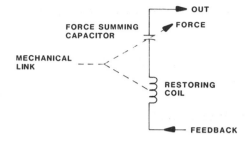

Figure 3–15   FORCE BALANCE
TRANSDUCER ELEMENT

Force Balance Transducer Elements

In Figure 3–15 the actual sensing element is the capacitor in the *force balance* element. However, this element could just as well be an inductive element. Operation of this device is rather simple. A change is effected by a measurand change and force is applied to a *force summing capacitor*. The physical properties of the capacitor change and therefore its capacitance also changes. The variable signal is routed to an amplifier and in turn to a servomechanism. A feedback signal equal to the variable output of the capacitor is fed back to the force balance element to return it to its original state before a measurand force change. The feedback may be mechanical and/or electrical and may be mechanized by a servo system or electromagnetism. Actual motion of the mechanical linkage and force summing devices is difficult to detect. The device has a fairly high output and is accurate and stable. It does, however, have a low-frequency response and is sensitive to acceleration and shock. The element is heavy and may be expensive.

Oscillator Transducer Elements

In Figure 3–16 a fixed capacitor and a *variable inductor* (coil) are used as a frequency-resonant circuit input to an oscillator. The variable unit may be either of these components. The *force element* may be, for example, water or air pressure. As the measurand changes, the force summing bar causes the inductor to change inductance, which alters the frequency at which the circuit is resonant.

Any change in the resonant circuit will be felt at the oscillator amplifier input:

$$\Delta f = f_o - f_r$$

where

$$\Delta f = \text{change in frequency}$$

$$f_o = \text{operating frequency}$$

$$f_r = \text{resonant frequency}$$

The *oscillator element* may be very small, with a fairly high output. Temperature may restrict the operating range and the devices used may play a part in poor

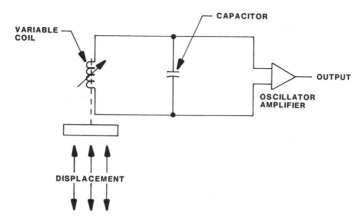

Figure 3–16   OSCILLATOR TRANSDUCER ELEMENT

sensitivity. Accuracy of the device may be poor unless expensive components are utilized.

Differential Transformer Transducer Elements

The *differential transformer* consists of primary coil, a pair of or several secondary windings, and a core (see Figure 3–17). The primary and secondary windings are inductors (coils) and as you would expect, have properties similar to the inductive transducers discussed previously. The core is moved by a force

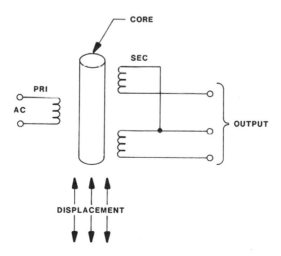

Figure 3–17   DIFFERENTIAL TRANSFORMER TRANSDUCER ELEMENT

from the measurand. In a majority of designs the core mass is a push rod of some description which is tied to linkage. The core moves between the primary and secondary coils as the measurand changes. Induced voltage between the primary and secondaries changes and an output is effected. The secondary output is usually fed to a demodulator or an ac bridge network.

The differential transformer transducer is best suited for large displacements. Signal output is larger than for most other transducers. The differential transformer is always driven by alternating current. The frequency of the ac current is usually 50 to 60 H from commercial power sources. Higher frequencies of 400 Hz to 10 kHz may be utilized to reduce sensor component sizes. Because of its large displacement, the differential transformer is susceptible to vibration. However, some of the modern linear models have overcome this problem. Again, all the factors that affect inductors and inductive transducers will also affect the differential transformer.

Photoelectric Transducer Elements

The photoelectric element consists of a *force summing diaphragm*, a window, a light source, and a detector (see Figure 3–18). Some measurand such as force or pressure causes the diaphragm to deflect. The window opens or closes as the diaphragm modulates the amount of light from the light source through the window opening. A photodiode on the window's opposite side detects the light and feeds the varying amounts as current flow to some receiving or amplifying component. The output then is dependent on the force applied by the measurand. This is a linear system and can be made extremely accurate with a strong and sensitive light-emitting diode (LED) as the light source. The unit is simple and can provide a high output. Frequency response may be low and long-term stability is not always achieved. Temperature, as in other transducer elements, is always a factor in efficiency.

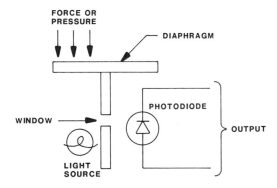

Figure 3–18 PHOTOELECTRIC FORCE SUMMING TRANSDUCER ELEMENT

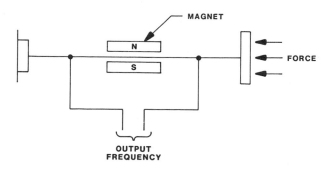

Figure 3–19 VIBRATING-WIRE TRANSDUCER ELEMENT

Vibrating-Wire Transducer Elements

The vibrating-wire element consists of a fine tungsten wire strung through a strong magnetic field so that the lines of force crossing the wire are maximum (see Figure 3–19). As the measurand causes a force change, the wire vibrates at a frequency that is determined by the length of the wire and tension applied. As the wire vibrates in the magnetic field, an electrical signal is generated at the output (taken from the wire) and fed to an amplifier. A feedback signal is fed back to the wire to maintain oscillation at the desired frequency. Modification of the wire-vibration frequency is then determined by the force applied. The output of the vibrating-wire element can be high and can be transmitted over long distances without much loss. The device is sensitive to acceleration and shock and is not considered a stable element. Temperature affects the wire and its hookup.

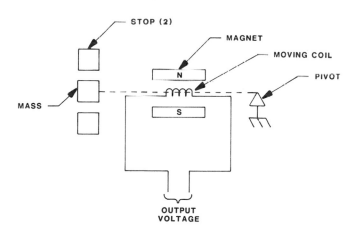

Figure 3–20 VELOCITY TRANSDUCER ELEMENT

Velocity Transducer Elements

The velocity element consists of a *moving coil* within a magnetic field (see Figure 3–20). One end of the coil is attached to a pivot, while the other end is free to move within a restricted area. Output is taken from the moving coil. The voltage signal is generated by the coil moving within the magnetic field. The output is proportional to the velocity of coil movement. Electrical feedback may be used for dampening purposes.

Strain Gage Transducer Elements

The purpose of a strain gage element is to detect the amount or length displaced by a force member. The strain gage produces a change in resistance that is proportional to this variation in length. Each strain gage has a property known as a *gage factor* which provides this function. Gage factor is a ratio of resistance and length:

$$GF = \frac{\Delta R/R}{\Delta L/L}$$

where

$$GF = \text{gage factor}$$

$$\Delta R = \text{change in resistance}$$

$$R = \text{original resistance}$$

$$\Delta L = \text{change in length}$$

$$L = \text{original length}$$

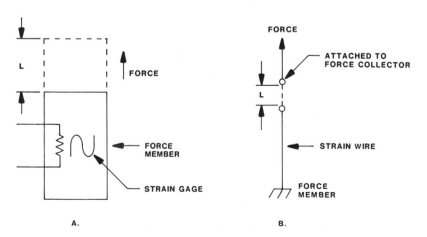

Figure 3–21   STRAIN GAGE TRANSDUCER ELEMENT

The gages are electrically installed as part of a Wheatstone bridge for circuit application.

There are two basic strain gage types: *bonded* and *unbonded*. The bonded gage (see Figure 3–21A) is entirely attached to the force member by adhesive. As the force member stretches in length, the strain gage also lengthens. The unbonded gage (see Figure 3–21B) has one end of its strain wire attached to the force member and the other end attached to a force collector. As the force member stretches, the strain wire also changes in length. Each motion of length either with bonded or unbonded gages causes a change in resistance. Strain gages are made of metal and semiconductor materials. Strain gages are very accurate, may be excited by alternating or direct current, and have excellent static and dynamic response. The signal out is small, but this disadvantage may be corrected with good periphery equipment.

Other Transducer Elements

There are certainly other specialized transducers made in this country and the rest of the world that do special tasks and have unique elements. We have attempted a coverage of most of today's transducer types.

4

Components
of Analog
Signal Conditioning

The *transducer* is a device that measures a dynamic variable and converts it into an electrical signal. That electrical signal is analogous to the variations of the dynamic variable and is termed an analog signal. *Analog*, then, may be described as an electrical representation of a physical quantity of data. *Analog data* are data that have a continuous set of values within a given range. For example, a temperature range of 0 to 250°F may be represented by an electric voltage of 0 to 10 V. As the temperature varies, so does the voltage. That voltage measured at any point in time is representative or analogous to some temperature in degrees Fahrenheit.

In a microprocessor-controlled system the analog signal, for all intents and purposes, is useless until it is converted to a form that is acceptable to the computer. An example of this is a transducer that outputs a variable microvolt signal (a low-level signal). This signal must be amplified to raise it to a high-level signal (say, 1 to 10 V) so that it may be converted to digital. The signal of 1 to 10 V (still analog) is then placed into an analog-to-digital (A/D) converter to interface with the computer bus. The entire process is called *signal conditioning*.

Other transducers produce high output voltages, frequency signals, pulses, low and high currents, noisy alternating signals, and so on. These must be amplified, divided, filtered, and otherwise massaged to convert them to a voltage level (say, 1 to 10 V) to make them compatible to the A/D converter.

Unfortunately, nature is not as simple as we would like. Therefore, each dynamic variable (physical phenomenon) must be treated as a single entity. We cannot use the same electronics to do the job. We must use combinations of components to do that job for us. In this chapter we provide those components of analog signal conditioning that are most used in industry today. In Chapter 5 we present those digital circuits utilized for conversion from/to the computer bus.

There are basically two types of analog signal conditioning, *passive* and *active*. *Passive analog signal conditioners* are those that are not modified by electronics. That is, they are *not* modulated, amplified, integrated, and so on. Passive analog signal conditioners utilize fixed discrete components such as the resistor, capacitor, and inductor. *Active analog signal conditioners* are those that modify the signals with the use of electronic devices, such as discrete solid-state circuits and integrated circuits.

The wide variety of transducers on the market provide an equally wide variety of variable electrical analogs. For the purposes of this book, "signal conditioning" refers to the electrical/electronic circuitry required to convert these analogs in a form that interfaces with the other components in the process control.

PASSIVE SIGNAL CONDITIONERS

Passive signal conditioners are limited to those circuits that utilize discrete components and do not modify the signal. As just stated, the passive conditioner uses resistors, capacitors, and inductors. It does not use solid-state devices or integrated circuits. Since this is the case, the circuitry is limited as to its capability. Process control utilizes the voltage divider and the bridge circuit extensively and the passive filter for elimination of high and low frequencies from the process.

Voltage Dividers

The *voltage divider* is the most fundamental of the signal conditioning types. It consists of a pair of resistances in series, one fixed and the other variable (see Figure 4–1). The resistance of the set resistance is chosen by some triggering requirement such as the output of a transducer. R_{VAR} is the varying transducer resistance. The transducer resistance decreases by an amount necessary to provide an output voltage that equals the voltage V_{SET}. The voltage divider is most useful when used with large varying resistances where linearity is not a problem. It is especially useful for triggering a level of a variable such as light or heat.

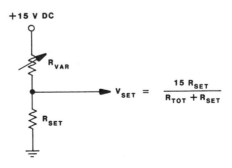

$$V_{SET} = \frac{15\,R_{SET}}{R_{TOT} + R_{SET}}$$

Figure 4–1 VOLTAGE DIVIDER

The Wheatstone Bridge Circuit

The primary measurement circuit used in instrumentation and specifically with strain gages is the resistance bridge. The prominent circuit type is the *Wheatstone bridge*. The elementary Wheatstone bridge is used to determine the resistance of any unknown resistance. This same bridge is used extensively in automatic control systems and with electrical measuring instruments.

The basic Wheatsone bridge is shown in Figure 4–2. R_1 and R_2 comprise a voltage divider from the power source. R_{UNK} (the unknown resistance) and R_{VAR} (the variable calibrated resistance) represent the second voltage divider from the power source. The R_1, R_2, and R_{VAR} resistances are all very accurate components. Their values are known.

A precision galvanometer is placed between points A and B. When power is applied to the circuit, the galvanometer will indicate current flow in either the point A or point B direction. Resistor R_{VAR} is then varied until the galvanometer indicates no current flow. This shows that the bridge is balanced. When it is balanced, the bridge sets up a mathematical voltage ratio between the two voltage dividers. This ratio is as follows:

$$\frac{I_1R_1}{I_1R_2} = \frac{I_2R_{UNK}}{I_2R_{VAR}}$$

The current component can be removed from the equation because current is equal in each leg. The ratio then reads

$$\frac{R_1}{R_2} = \frac{R_{UNK}}{R_{VAR}} \qquad \text{or} \qquad R_{UNK} = R_{VAR} \times \frac{R_1}{R_2}$$

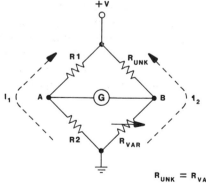

$$R_{UNK} = R_{VAR} \times \frac{R1}{R2} \qquad \textbf{Figure 4–2} \quad \text{WHEATSTONE BRIDGE}$$

The Wheatstone Bridge Used in Signal Measurement

In the search for accuracy, the Wheatstone bridge seems to stand tall for measurement of small resistance changes. Figure 4–3 is representative of a balanced bridge circuit. The terminals of the output (E_o) are in balance and have a difference

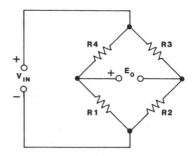

Figure 4–3 WHEATSTONE BRIDGE
USED IN SIGNAL MEASUREMENT

AT BALANCE, $E_0 = 0$ IF $\frac{R1}{R4} = \frac{R2}{R3}$

of potential of zero. The output E_o is zero. A bridge is used for three reasons: as a circuit for determining an unknown resistance, as a null detector, and as a circuit to measure the difference between two voltages or current.

There is a common resistance ratio for the bridge shown in Figure 4–3. This ratio is

$$\frac{R_1}{R_4} = \frac{R_2}{R_3}$$

If three resistances of this ratio are known, the other can easily be calculated. For example, if the resistance R_3 is unknown, then

$$R_3 = \frac{R_2 R_4}{R_1}$$

As a null detector, the output of the bridge would be at null ($E_o = 0$ V) when and if the bridge were balanced.

As a detector of voltage the input V_{IN} is placed between opposite terminals, while the output E_o is taken from a second set of terminals. In essence, the bridge output is measuring the difference voltage between the leg pairs. The first leg is a series voltage divider looking at the voltage across resistor R_1.

$$V_{R1} = \frac{R_1}{R_1 + R_4} V_{IN}$$

The second leg is a series voltage divider looking at the voltage across resistor R_2.

$$V_{R2} = \frac{R_2}{R_2 + R_3} V_{IN}$$

The calculation of E_o is the difference voltage between the two series legs.

$$E_o = \frac{R_1}{R_1 + R_4} V_{IN} - \frac{R_2}{R_2 + R_3} V_{IN}$$

or

$$E_o = \frac{R_1/R_4 - R_2/R_3}{(1 + R_1/R_4)(1 + R_2/R_3)} V_{IN}$$

The bridge as shown in Figure 4–3 allows for the detection and measurement of small resistance changes. A change from bridge balance may easily be monitored. The output E_o changes as any adjacent leg resistance changes. The change is represented by a difference of potential between the parallel legs (at E_o). There also may be cases when adjacent legs change resistance by the same amount. In this event, the bridge would still be balanced $R_1/R_4 = R_2/R_3$. The output voltage (E_o) would remain at a null of zero.

In transducer arrrangements, the ratio of fixed resistances such as R_2/R_3 is called ratio K. Let us assume that R_2 and R_3 are fixed at a *constant ratio K*. If R_1 is unknown and R_4 is an accurate variable resistance, the size of R_1 can be determined by adjusting R_4 until a null is achieved. The null will occur when $R_1/R_4 = R_2/R_3$. If R_4 is fixed, a null will occur when R_1 becomes equal to K times R_4.

Passive Filters

The final circuit type that is used extensively for passive signal conditioning applications is the *passive filter*. The passive filter may be a simple circuit constructed of a capacitor and a resistor (see Figure 4–4).

The *low-pass filter* is a series *RC* circuit with the capacitor in parallel with the output. This circuit operates as its name implies, by allowing low-frequency signals to pass while rejecting high-frequency signals. During control of a process,

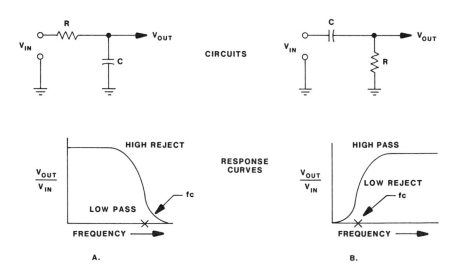

Figure 4–4   PASSIVE FILTERS: (A) LOW-PASS FILTER, (B) HIGH-PASS FILTER

the computer and its digital circuits, such as gates, latches, and multivibrators, produce switching spikes which cause undesirable frequency noise. Low-pass filters are utilized to attenuate the high-frequency noise and allow the lower frequencies to pass.

The low-pass filter circuit will attenuate all frequencies near and above a critical (chosen) frequency. The reader is referred to the following attenuation equation:

$$f_c = \frac{1}{2\pi RC}$$

where

f_c = critical (chosen) frequency

R = resistance

C = capacitance

The curve for the low-pass filter is simplified for quicker understanding. The response of voltage output to voltage input against frequency is plotted as a logarithm. Low frequencies are passed, while high frequencies are attenuated gradually about the critical frequency f_c. It is possible to cascade the filters to demand more attenuation.

The *high pass filter* is a series RC circuit with the resistance in parallel with the output. This circuit operates as the name implies, by rejecting low-frequency signals while allowing high-frequency signals to pass. During control of a process, the computer may experience some low-frequency noise, usually from the ac power line. High-pass filters are utilized to attenuate the low-frequency noise and allow the higher frequencies to pass.

The high-pass filter circuit will attenuate all frequencies below the critical (chosen) frequency. Use of the attenuation equation discussed previously is the same. The curve for the high-pass filter is simplified for quicker understanding. The response of voltage output to voltage input against frequency is plotted as a logarithm. High frequencies are passed while low frequencies are attenuated gradually about the critical frequency (f_c). Cascading may be used to improve attenuation up to the point of circuit loading.

These low- and high-pass filter circuits are extremely basic. They are only meant to represent passive filters. It is not often that a circuit of this simplified level may be used. Reference is made to most digital technology textbooks.

ACTIVE SIGNAL CONDITIONERS

The active signal conditioner utilizes special solid-state devices or integrated circuits to modify the transducer (or other) signals. The active circuit may amplify, attenuate, modulate, integrate, differentiate, and so on, to condition the signal for use in the control process. The use of special devices distinguishes the active circuits from those that use resistor and capacitors in passive networks. In this section we discuss just a few of the circuits that are used extensively for active signal conditioning.

SOLID-STATE DEVICES IN DISCRETE OPERATION

Bipolar Transistors

The bipolar transistor is a two-junction, three-layer, solid-state device constructed on a single crystal (see Figure 4–5). It has two basic functions, switching and amplification. Switching functions are used in digital circuitry. Amplification is performed in linear and/or analog circuitry. The transistor uses both N- and P-type silicon. A thin layer of P-type material is sandwiched between thick layers of N-type material to form an NPN transistor. A thin layer of N-type material is sandwiched between thick layers of P-type material to form a PNP transistor.

The transistor has three terminals. The thick layers on the ends are *collector* and *emitter*. The thin layer in the center is the *base*. The base is extremely thin, about 1 micrometer (μm) or smaller. The shape of the transistor package is usually dependent on the amount of current flow it can handle, together with the application.

The transistor has for many years been the mainstay of the electronic field for amplifying. Since the development of the operational amplifier integrated circuit, the transistor amplifier has taken a far second place in amplification. The author believes, however, that the solid-state discrete amplifier will be around for several more years. Therefore, some basic transistor amplifier fundamentals should be explained.

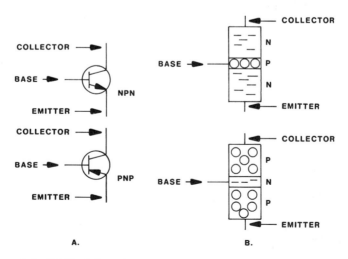

Figure 4–5   BIPOLAR TRANSISTOR: (A) SYMBOLS, (B) JUNCTION MODELS

Common-Emitter Amplifier Configuration

The common-emitter (CE) amplifier configuration is the most used of the amplifiers because it has high voltage and current gain (therefore, high power gain). The CE amplifier has a phase reversal between input signal and output signal. For this reason, two stages are often used. If phase is not a problem, the configuration is ideal.

Figure 4–6 is a typical common-emitter amplifier. The input signal is fed into the base in reference to the emitter. The output signal is taken off the collector in reference to the emitter.

Let's consider the dc circuit analysis first. Positive V_{CC} supplies reverse bias to the collector. The point Y in reference to ground is approximately one-half the dc voltage V_{CC}. The dc voltage drops on the output leg of the amplifier sum up to applied voltage:

$$V_{CC} = V_{RL} + V_{CE} + V_{RE}$$

R_{B1} and R_{B2} together are a voltage divider for forward bias on the base–emitter junction. Bias voltage V_B is equal to the base–emitter junction voltage V_{RBE}, plus the emitter voltage V_{RE}. Bias voltage V_B can also be calculated using the voltage-divider formula in the following manner:

$$V_B = V_{CC} \frac{R_{B2}}{R_{B1} + R_{B2}}$$

The base–emitter junction voltage is approximately 0.6 V dc if the transistor is made of silicon and 0.2 V dc if the transistor is made of germanium. These values are approximations and are dependent on and vary with temperature.

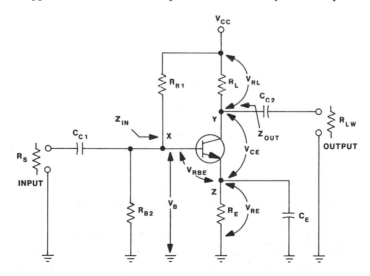

Figure 4–6 COMMON-EMITTER (CE) AMPLIFIER CONFIGURATION

The coupling capacitors C_{C1} and C_{C2} are charged to dc voltages at points X and Y, respectively. The signal is placed on the input X in relation to ac ground. Since the emitter resistor R_E is bypassed by the capacitor C_E, the point Z is essentially ac ground. This means that the signal will be felt only across the base–emitter junction. Junction dynamic resistance is calculated as an approximation using the following formula:

$$r_{be} = \frac{25}{I_E}$$

where

$$25 = \text{Shockley constant}$$

$$I_E = \text{dc operating current, milliamperes}$$

This formula was developed by William Shockley to present an estimate of dynamic base–emitter resistance for analysis purposes.

Once the dynamic resistance r_{be} of the base–emitter junction has been estimated, the input impedance can be calculated. Input impedance is an arrangement of parallel paths to ac ground looking into the amplifier at point X.

$$Z_{\text{in}} = R_{B1} \| R_{B2} \| \beta r_{be}$$

where β is the current gain of the transistor (beta). Output impedance Z_{out} is the ac ground path looking back into the amplifier at point Y. Since there is but one path to ground for ac (signal), the output impedance is simply equal to R_L.

The voltage gain A_v of the amplifier is a ratio of output signal to input signal. This first formula is applicable for the voltage gain with a working load.

$$A_v = \frac{R_l\,\text{ac}}{r_{be}}$$

where

$$R_l\,\text{ac} = R_L \text{ in parallel with } R_{LW}$$

$$R_{LW} = \text{working load resistance}$$

This second formula is used for calculating voltage gain without a working load.

$$A_v = \frac{R_L}{r_{be}}$$

Common-Emitter Amplifier without a Bypass Capacitor

Without a bypass capacitor, the CE amplifier is a low-voltage gain amplifier (see Figure 4–6). The voltage gain of this amplifier is again the ratio of the input signal to the output signal. However, the input signal is felt across the base–emitter junction and the emitter resistor.

$$Av = \frac{R_l \, ac}{r_{be} + R_E}$$

where

$$R_l \, ac = R_L \text{ in parallel with } R_{LW}$$

$$R_{LW} = \text{working load resistance}$$

The input impedance also changes somewhat from the bypassed CE amplifier. Considering that the bypass capacitor has been removed, the ac path to ground is across the emitter resistor R_E.

$$Z_{in} = R_{B1} \| R_{B2} \| \beta(r_{be} + R_E)$$

Common-Collector Amplifier Configuration

The common-collector (CC) amplifier configuration, also called the emitter follower (Figure 4–7), is used for impedance matching and for voltage gains of less than unity (1). It has good current gain, similar to the common emitter. The CC amplifier has no phase reversal between input and output signal. This is, of course, a good point if phase is a problem.

Figure 4–7 is a typical common-collector amplifier. The input signal is fed into the base in reference to ac ground. The output is taken off the emitter resistor. This is why the configuration is often called an emitter follower.

Let's consider the dc circuit analysis first. Positive V_{CC} supplies reverse bias to the collector. The dc voltage drops on the output leg of the amplifier sum up to applied voltage.

$$V_{CC} = V_{CE} + V_{RE}$$

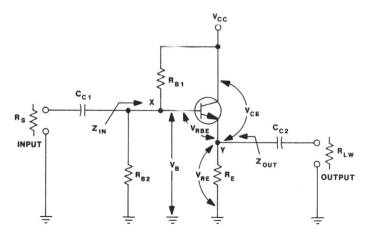

Figure 4–7   COMMON-COLLECTOR (CC) AMPLIFIER CONFIGURATION

R_{B1} and R_{B2} are a voltage divider for forward bias on the base–emitter junction.

Bias voltage V_B is equal to the base–emitter junction voltage V_{RBE} plus the emitter voltage V_{RE}. Bias voltage V_B can also be calculated using the voltage-divider formula in the following manner:

$$V_B = V_{CC} \frac{R_B R_{B1}}{R_{B1} + R_{B2}}$$

The base–emitter junction voltage is approximately 0.6 V dc if the transistor is made of silicon and 0.2 V dc if the transistor is made of germanium. These values are approximations and are dependent on, and vary with, temperature.

The coupling capacitors C_{C1} and C_{C2} are charged to dc voltages at points X and Y, respectively. The signal is placed on the input at point X in relation to ac ground. The collector is at ac ground. Since the output is in parallel with the emitter resistor, this signal is essentially the same as the input, less the base–emitter dynamic resistance r_{be}.

Junction dynamic resistance (r_{be}) is calculated as an approximation using the following formula:

$$r_{be} = \frac{25}{I_E}$$

where

$$25 = \text{Shockley constant}$$

$$I_E = \text{dc operating current, milliamperes}$$

Once the dynamic resistance r_{be} of the base–emitter junction has been estimated, the input impedance can be calculated. Input impedance is a parallel path to ac ground looking into the amplifier at point X.

$$Z_{in} = R_{B1} \| R_{B2} \| \beta r_{be}$$

where

$$\beta = \text{current gain of the transistor}$$

$$r_{be} = R_E \| R_{LW}$$

Output impedance Z_{out} is the ac ground path looking back into the amplifier at point Y.

$$Z_{out} = R_E \| r_{be} + \frac{R_{B1} \| R_{B2} \| R_s}{\beta}$$

where R_S is the input source resistance.

The voltage gain A_v of the amplifier is a ratio of the output signal to the input signal.

$$A_v < 1$$

The voltage gain is actually less than unity. All the signal input is felt across the emitter resistor R_E and the parallel working load resistor R_{LW}, except for a minor amount dropped across the base–emitter junction.

Common-Base Amplifier Configuration

The common-base (CB) amplifier configuration (Figure 4–8) has a high voltage gain but a current gain of less than unity (1). Current gain in a common-base amplifier is called alpha (α). *Alpha* is simply a ratio of input current to output current. The signal is applied to the emitter and taken from the collector. Therefore,

$$\text{alpha } (\alpha) = \frac{I_C}{I_E}$$

Since the emitter current includes the collector current, the current gain must be less than unity (1).

Alpha (α) may be converted to the more useful *beta* (β) with the following equation:

$$\beta = \frac{\alpha}{1 - \alpha}$$

The CB amplifier has no phase reversal between input signal and output signal. Figure 4–8 is a typical common-base amplifier. We have used a *PNP*-type transistor in this configuration. Therefore, a negative power supply is required. The input signal is fed into the emitter in reference to the base. The base is coupled to ac ground with capacitor C_B. The output signal is taken off the collector in reference to the base.

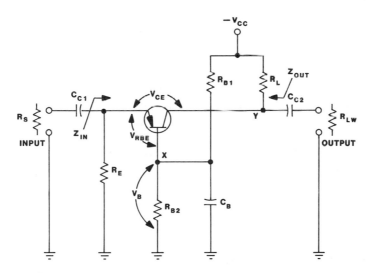

Figure 4–8 COMMON-BASE (CB) AMPLIFIER CONFIGURATION

Let's consider the dc circuit analysis first. Negative V_{CC} supplies reverse bias to the collector. The point Y in reference to ground is approximately one-half the dc voltage $-V_{CC}$. V_{CE} is the collector–emitter dc voltage operating range.

The resistors R_{B1} and R_{B2} are a voltage divider for forward bias on the base–emitter junction. Bias voltage V_B is equal to the base–emitter junction voltage plus the voltage drop across emitter resistor R_E. Bias voltage can also be calculated using the voltage-divider formula:

$$V_B = -V_{CC} \frac{R_{B2}}{R_{B1} + R_{B2}}$$

The base–emitter junction voltage is approximately 0.6 V dc if the transistor is made of silicon and 0.2 V dc if the transistor is made of germanium. These values are approximations and are dependent on and vary with temperature.

C_{C1} and C_{C2} are capacitors that couple the input signals into the amplifier and out of the amplifier. The signal is placed on the emitter resistor, which is in parallel with the base–emitter junction r_{be}. The base is at ac ground. The junction dynamic resistance r_{be} can be calculated using the following formula:

$$r_{be} = \frac{25}{I_E}$$

where

$$25 = \text{Shockley constant}$$

$$I_E = \text{dc operating current, milliamperes}$$

Once the dynamic resistance r_{be} of the base–emitter junction has been estimated, the input impedance can be calculated. Input impedance sees two parallel paths to ac ground.

$$Z_{\text{in}} = R_E \| r_{be}$$

and

$$Z_{\text{in}} \simeq r_{be} \text{ (because } r_{be} \text{ is so small)}$$

Output impedance Z_{out} is the ac ground path looking back into the amplifier at point Y. Since there is but one path to ground for ac (signal), the output impedance is simply equal to R_L.

The voltage gain of the amplifier is a ratio of output signal to input signal. The following formula is applicable for the voltage gain with a working load.

$$A_v = \frac{R_l \text{ ac}}{r_{be}}$$

where

$$R_l \text{ ac} = R_L \text{ in parallel with } R_{LW}$$
$$R_{LW} = \text{working load resistance}$$

A second formula is used for calculating voltage gain without a working load.

$$A_v = \frac{R_L}{r_{be}}$$

Unijunction Transistors

The unijunction transistor is a two-layer, three-terminal device. It is used as a relaxation oscillator and as a timing device to trigger SCRs. The unijunction is not an amplifier but does have large current gain.

The symbol for the UJT is illustrated in Figure 4–9A. The junction model is shown in Figure 4–9B. Note that the model has a comparatively large bar of N-type material and a small P-type section forming the junction. In operation, a positive potential is applied to base 2 and a negative to base 1. Very little current flows because of the resistance of the bar of N-type material. When a positive potential is applied to the emitter, it forward biases the junction of the emitter and base 1. This essentially decreases the resistance between the emitter and base 1 and, in turn, the resistance between base 1 and 2. A lowering of resistance in the N-bar increases the current flow from base 1 to base 2. More emitter voltage causes more current flow through the bases.

In Figure 4–9C a characteristic curve of the operation is shown. When voltage is applied to the emitter in a large enough forward (positive) direction, it will overcome a reverse bias provided by the base 1 and 2 potentials. The resistance in the N material sets up a voltage divider, which reverse biases the junction. The resistance ratio is called the *intrinsic standoff ratio*. It is this action that allows the UJT to be triggered. The UJT is packaged the same as a bipolar transistor.

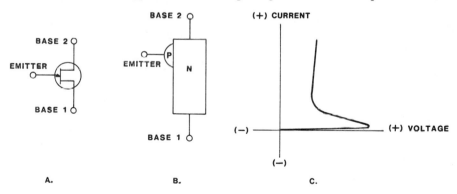

Figure 4–9   UNIJUNCTION TRANSISTOR: (A) SYMBOL, (B) JUNCTION MODEL, (C) CHARACTERISTIC CURVE

Unifunction Applications

UJT is used as either an oscillator or a trigger. In Figure 4–10A, the UJT is applied as a *relaxation oscillator*. Power is applied to the circuit. The capacitor C_1 begins to charge through the variable resistance R_v. The UJT is biased (+) on

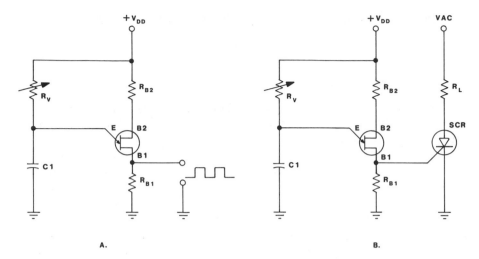

Figure 4–10 UNIJUNCTION APPLICATIONS: (A) RELAXATION OSILLATOR, (B) TRIGGERING

base 2 and (−) on base 1. The junction between the emitter and the bases is reverse biased. There is no conduction. As the capacitor reaches a predesignated charge, current flows from base 1 to base 2. A second current path develops when the capacitor discharges through R_{B1} and the emitter–base 1 junction. The output is a pulse across resistor R_{B1}. A similar wave form may also be taken from the terminal at base 2. This pulse would be 180° out of phase with the waveform taken from base 1. The capacitor discharges to some level below triggering, and the cycle begins again.

The second application (Figure 4–10B) is triggering. In this case the UJT is used as a trigger for a silicon-controlled rectifier (SCR). In the circuit, the relaxation oscillator just explained is used for triggering. The frequency of the output of the oscillator is dependent on the values of R_V and C_1. The pulse turns on the SCR only when its anode is positive. Current is applied through the load. You may recall that the SCR, when turned on, will not turn off until power is removed, regardless of its trigger condition. Since the power source is alternating, the SCR will be off during negative alternations. It will be on when the pulse is positive and the UJT triggers its gate.

Junction Field-Effect Transistors

The *junction field-effect transistor* (JFET) is a simple amplifying device made of a sandwich on *N*- and *P*-type materials. The JFET symbol is shown in Figure 4–11A. Its junction model is illustrated in Figure 4–11B. The JFET is used for circuitry requiring very large input impedances. For example, the integrated-circuit operational amplifier often has a first stage made of a pair of JFET circuits. The JFET may also be used as a high-frequency amplifier and a switch.

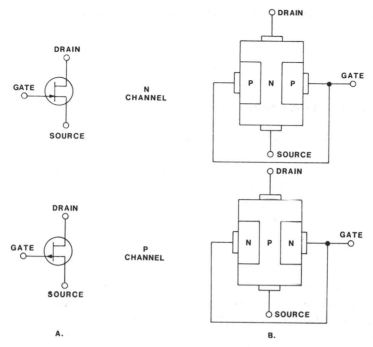

Figure 4-11 JUNCTION FIELD-EFFECT TRANSISTOR (JFET): (A) SYMBOLS, (B) JUNCTION MODELS

The junction model in Figure 4-11B presents JFET operation. Let's consider the N-channel JFET in the upper section of the figure. The N material forms the channel, which extends between two terminals called the *drain* (D) and the *source* (S). Positive voltage polarity is applied to the drain and negative to the source. A slab of P material is diffused on both sides of the channel to form a sandwich-shaped structure.

In a no-bias condition (zero bias) on the gate, there is no depletion area at the junction of the gates and the channel. Current flows freely from source to drain. If reverse bias is applied to the gates (−), a depletion area is expanded from the junction. The larger the (−) is, the larger the depletion area. If the (−) is large enough, this area will spread across the channel and stop current flowing from source to drain. This condition is called *pinch-off*. Reverse bias depletes the channel of carriers as in a reverse-biased diode. The extreme conditions are zero bias and pinch-off. All areas in between these two conditions provide variable current from source to drain. It is in this variable range that amplification takes place.

In the lower section of Figure 4-11, the P-channel JFET is modeled. This JFET operates in the same manner as the N-channel type. The exception is that pinch-off reverse bias on the gate is (+) and amplification takes place between zero and (+) pinch-off voltage.

The *order of merit* (amplification factor) is a ratio of source to drain current and gate voltage:

$$g_m = \frac{\Delta I_D}{\Delta V_G}$$

The JFET As An Amplifier

The circuit shown in Figure 4–12 is typical of the operation of a JFET in an application as a practical amplifier. This particular amplifier is called a common-source amplifier and is used throughout industry. It may be related to the common-emitter amplifier using bipolar transistors (see Figure 4–6). The JFET shown is an *N*-channel type. The resistors R_{G1} and R_{G2} form a voltage divider that sets the gate voltage (V_g) as follows:

$$V_G = V_{CC} \frac{R_{G2}}{R_{G1} + R_{G2}}$$

Circuit input impedance is calculated by looking at the resistance between ground and the gate terminal. This consists of the voltage-divider resistances and the input resistance of the JFET. Since the input resistance of the JFET itself is infinite, we may ignore that value.

Z_{IN} may be calculated as follows:

$$Z_{IN} = R_{G1} \| R_{G2}$$

where

$$Z_{IN} = \text{input impedance}$$

$$R_{G1} \text{ and } R_{G2} = \text{voltage-divider resistance}$$

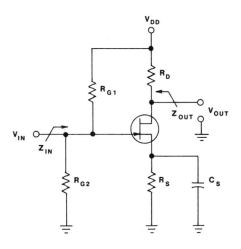

Figure 4–12 JFET AS AN AMPLIFIER

Output impedance is essentially the drain resistance. Output resistance has very little effect on the total output impedance. Therefore,

$$Z_{out} = R_D$$

where

$$Z_{out} = \text{output impedance}$$

$$R_D = \text{drain resistance}$$

The voltage gain of the amplifier, as with all amplifiers, is a ratio of voltage out to voltage in.

$$A_v = \frac{V_{out}}{V_{in}} \quad \text{(voltage gain)}$$

We must clarify this further. The *mutual inductance* (figure of merit) of the circuit, as you may recall, is calculated in the following manner:

$$g_m = \frac{\Delta I_D}{\Delta V_G}$$

If we combine the output impedance and transconductance equations, we create a voltage-gain formula for the voltage change between the gate and the drain.

$$A_v = g_m R_D$$

The reader must realize that the mutual conductance and gain formulas are basic and would be modified with additional circuitry. Additions of capacitors and resistance in the bias legs or the source and drain circuits could modify these basic equations. If the source resistor is not bypassed as in Figure 4–12, the effect is a lowering of voltage gain. It will, however, provide greater bandwidth.

The JFET As A Source Follower

As with the *bipolar transistor* amplifier, the JFET may be used as an isolation amplifier because of its high input impedance. The source follower is similar to the emitter follower of the bipolar transistor because it also has a very low output impedance. In Figure 4–13 the JFET is used as a *source follower*. It is called a source follower because the output is taken from the source resistance R_S. The circuit is also called a *common-drain* configuration because the drain is connected directly to the dc supply and therefore at ac ground.

The resistors R_{G1} and R_{G2} provide a voltage divider that sets the dc gate voltage.

$$V_G = V_{CC} \frac{R_{G2}}{R_{G1} + R_{G2}}$$

The input signal V_{in} sees the gate-to-source resistance $1/g_m$ and Z_{IN}. The output signal V_{out} is taken across the source resistor. The input signal V_{in} physically

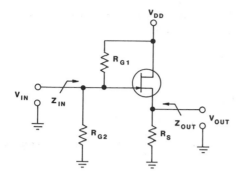

Figure 4–13   JFET AS A SOURCE FOL-
LOWER

appears from the gate to ground across the source resistor R_S. The output signal V_{out} essentially sees the source resistor R_S. The only difference between input and output voltages is the small voltage drop across the gate-to-source resistance r_s $(1/g_m)$ of the JFET device.

It stands, then, that input and output voltages are approximately the same:

$$V_{in} \simeq V_{out}$$

The voltage gain is

$$A_v \simeq 1$$

Input impedance Z_{IN} is the same as in the common-source configuration.

Output impedance is extremely low because we are looking back at the circuit between the source terminal and ground. This sees the source resistance R_S in parallel to the gate-to-source dynamic resistance r_s $(1/g_m)$:

$$Z_{OUT} = R_S \parallel r_s$$

The outcome is a very low output impedance.

Metal-Oxide Semiconductor Field-Effect Transistors

Another FET device is the *metal-oxide semiconductor field-effect transistor*, or *MOSFET*. The MOSFET differs only slightly from the JFET, described earlier. The principle of operation is the same as the JFET except that the gate is made of metal, insulated by a thin layer of glass or oxide. Operation is by current conduction, and control is achieved by varying an electric field.

There are two types of MOSFET: *depletion* and *enhancement*. The depletion type conducts with the gate at zero bias. Conduction increases by applying further negative bias until cutoff. The enhancement type is cut off at zero bias, whereas further positive bias allows it to conduct.

MOSFETs are used in digital logic and memory circuits. Furthermore, MOSFETs are used in op amps and other linear circuits but are not as popular as the bipolar transistor for linear work. The MOSFET is popular for logic circuits as it does not require resistors or diodes. Because of this, integrated circuits may be extremely small. Furthermore, they are simple to manufacture. For these reasons, the MOSFET has become the workhorse of very large scale integration (VLSI) circuits, which include devices such as microprocessors. MOSFETs are packaged in typical bipolar transistor cases.

Depletion-Type MOSFETs

In Figure 4–14A the *depletion-type MOSFET* contains a *P* substrate, an *N* channel, and a metal gate that is insulated from the semiconductor material with an oxide. Source and drain terminals are connected to the drain and source diffused material. This is an *N*-channel depletion MOSFET. Its symbol is just above the *P*-channel depletion MOSFET symbol in Figure 4–14A. The structural model of the *P*-channel depletion MOSFET is the same as the *N* channel with the substrate made of *N* material and the channel of *P* material.

In operation the depletion MOSFET is much like the JFET. Two heavily doped wells provide a low-resistance connection between the ends of the *N* channel and the source and drain connections. The *N* channel is formed on the *P* substrate. Oxide is grown on the *N*-channel surface. Metal is deposited on the oxide to form the gate.

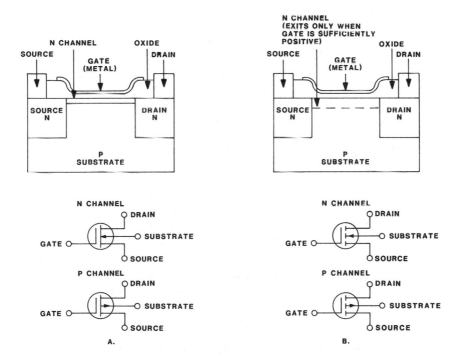

Figure 4–14 METAL-OXIDE SEMICONDUCTOR FIELD-EFFECT TRANSISTOR (MOSFET): (A) DEPLETION MOSFET, (B) ENHANCEMENT MOSFET

With zero bias, the electrons flow freely through the N channel from source to drain. As negative bias is increased, electrons are attracted by holes in the substrate, making them unavailable for conduction. A depletion area will result just below the oxide and will grow larger as negative bias is increased. The result is that the N channel will get thinner. Finally, current will be pinched off from source to drain.

Enhancement-Type MOSFETs

In Figure 4–14B the *enhancement-type MOSFET* contains a P substrate and a metal gate that is insulated from the semiconductor material with an oxide. Source and drain terminals are connected to the drain and source diffused materials. This is an N-channel enhancement MOSFET. Its symbol is just above the P-channel enhancement MOSFET symbol in Figure 4–14B. The structural model of the P-channel enhancement MOSFET is the same as the N channel with substrate made of N material and the channel of P material.

In operation the enhancement MOSFET is much like the JFET. Two heavily doped wells provide a low-resistance connection between the ends of the N channel, which exists when the gate is sufficiently positive. Oxide is grown on the top side of the P substrate. Metal is deposited on the oxide to form the gate.

You should be able to see that the physical difference between depletion and enhancement MOSFETs is that there is no channel. The channel does not exist if there is no bias. The current path from source to drain is as two diodes in series back to back. No current flows between source and drain without bias.

If a positive voltage (bias) is applied to the gate, it will attract electrons from the substrate and create an N-channel just below the oxide. This channel will become large enough to allow current to flow when a threshold voltage (specified) has been reached. As positive bias is increased, drain current increases.

MOSFETs and Electrostatic Charges

The MOSFET has extremely low leakage current and extremely high input resistance. It is easily destroyed because of these parameters. If a large voltage is applied from the gate to the channel, the oxide will rupture due to electrostatic field stress. This voltage can appear simply by electrostatic discharge from handling. Care must be taken during all handling tasks. When shipping the device, the leads are shorted together. When being handled, the tools and soldering irons are grounded. Technicians even use grounding straps around their wrists to prevent accidental electrostatic voltage applications between the gate and the channel.

The MOSFET As An Amplifier

There are many MOSFET amplifier configurations. They cannot all be covered here. The amplifier in Figure 4–15 is typical of a universal circuit using the enhancement MOSFET.

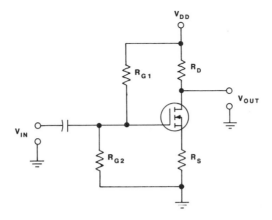

Figure 4–15 MOSFET AS AN AMPLIFIER

Our analysis begins with the voltage-divider resistors R_{G1} and R_{G2}. The dc voltage drops across these resistors may be calculated easily using the voltage-divider formula for either.

$$V_{RG1} = V_{DD} \frac{R_{G1}}{R_{G1} + R_{G2}}$$

$$V_{RG2} = V_{DD} \frac{R_{G2}}{R_{G1} + R_{G2}}$$

Operating drain current is determined by use of the device characteristic curves. Operating drain current is calculated using the following equation.

$$I_{DO} = \frac{I_D}{[1 - (V_{GS}/V_T)]^2}$$

where

I_{DO} = operating (saturation) current of the device

I_D = drain current for specified gate voltage

V_{GS} = gate voltage

V_T = threshold voltage

The voltage drops across the source resistor and the drain resistor are dependent on the operating current.

$$V_{RS} = I_{DO} \times R_S \qquad \text{and} \qquad V_{RD} = I_{DO} \times R_D$$

The figure of merit (voltage-gain factor) of the MOSFET amplifier is mutual conductance (transconductance). This value must be decided so that it may be used for further voltage-gain calculations:

$$g_m = \frac{2}{V_T} \sqrt{I_{DO}I_D} \quad \text{and} \quad A_v = g_m R_D$$

where

A_v = voltage gain

g_m = mutual conductance

R_D = drain resistance

V_T = threshold voltage

I_{DO} = operating current (saturation)

I_D = drain current at specified gate-to-source voltage (V_{GS})

These equations are very general in form and should be considered as such. If the reader should require in-depth use of the MOSFET rather than generalization of functions, it would be useful to acquire a major book on the subject.

INTEGRATED CIRCUITS

Integrated-Circuit Arrays

A simple and popular linear integrated circuit (IC) is called the *array*. It is several semiconductors on a single IC. The difference between this and other ICs is that the several circuits in the IC may or may not perform a specific function. They may be hooked up in any manner that the designer sees fit. For example, several diodes and transistors may be on one array. Others may have two differential amplifiers. Still others may have a bridge rectifier and other separate diodes. Linear ICs have evolved from the use of transistors and MOS devices in separate devices to combinations that include both these fabrication techniques. The usual technique in combinations is to use a FET (*field-effect transistor*) at the input. This supplies the device with an extremely low input current, a notable attribute.

Linear Integrated Circuits

By far the predominant linear integrated circuit is the operational amplifier. In fact, this device is used so often in the electronics industry that many books have been written about it. As you can see by the following list, there is a fairly large variety of linear integrated circuits. Linear integrated circuits perform process and control operations, signal generation, and amplification. The major use is with operational amplifiers. However, the voltage regulators and comparators are in increasing use.

1. Operational amplifiers
2. Voltage regulators

3. Voltage references
4. Instrumentation amplifiers
5. Voltage comparators
6. Analog switches
7. Sample-and-hold circuits
8. Analog-to-digital (A/D) and digital-to-analog (D/A) conversion
9. Telecommunications circuits
10. Audio/radio/television circuits
11. Transistor/diode arrays
12. Special function circuits

Confusion may arise from the word "linear." Linear in electronic devices means that cause and effect are proportional for all values of input and output. Natural devices in real life are not linear. In fact, the linearity across a range of operations is limited. Devices can, however, be nonlinear in operation but still have a useful life because the nonlinearity can be well defined, controllable, stable, and available at low cost. Analog relationships such as multiplication, square laws, and log ratios are examples of usefulness. Applications include modulation, power measurement, signal shaping, and correcting for the nonlinearity of measuring devices.

In general, the nonlinear device may be classified according to its smoothness. If the functions are smooth, it may be classified as *continuous function*. If it has one or more discontinuities or jumps, it is classified as a *discontinuous nonlinear function*.

Introduction to Operational Amplifiers

An operational amplifier IC is a solid-state integrated circuit that uses external feedback to control its functions. The operational amplifier may be used for a great number of circuits, which include amplifiers, attenuators, integrators, filters, differentiators, voltage followers, oscillators, regulators, and mathematical circuits. Its greatest characteristic is simplicity in operation.

The operational amplifier can replace entire transistor amplifier circuits in many applications, such as audio, radio, analog, and signal-conditioning systems. Its basic operation depends on the use of negative feedback, as described later.

Each operational amplifier has unique qualities that can be determined from the manufacturer's specifications. Some of these qualities are wide frequency range, internal frequency compensation, low power consumption, internal "short" protection, and temperature stability. Operational amplifiers are, in fact, the integrated-circuit evolution to the differential amplifier. They are especially noted for their high input impedance, high voltage gain, low output impedance, and zero output signal for zero input signal (low offset).

Our description of the operational amplifier will be brief because of the nature of this book. For a complete description of the op amp and its operation, see the author's *Operational Amplifiers* (Reston Publishing Co., Reston, Virginia, 1983).

Op Amp Principles of Operation

The principle of feedback in the operational amplifier IC is the function that makes it so versatile (see Figure 4–16). This principle is quite simple. A feedback connection is made between the output and inverted input leads, with resistors used as voltage dividers. The voltage divider provides the exact amount of negative feedback for any level of amplification (gain) required or desired.

Closed-loop gain is dependent on the size of the feedback resistance. *Open-loop gain* (no feedback resistance) can be ignored in most designs, but is used in voltage-comparison and level-detection circuits. The gain of the operational amplifier is controlled externally. The operational amplifier is said to have infinite gain within the restraints of power supplied at pins 4 and 7. It is also said to have infinite input impedance and zero output impedance. When there is no signal applied to either the inverting (pin 2) or noninverting (pin 3) inputs, the output (pin 6) is zero. When a signal is applied to one input, the opposite input strives to approach that signal level so that there is always a differential of zero between the two inputs.

Since *feedback* is from output to *inverting input*, the output will always be that amount necessary to drive the two inputs to near null (or an extremely small value). Current is so small at the inputs that it is assumed to be nonexistent. Output polarity is in phase with the *noninverting input* and out of phase with the inverting input.

Differences Between Operational Amplifiers

Each operational amplifier is constructed to perform at its best under certain conditions. Families of operational amplifiers are designed to optimize a set of parameters crucial to particular applications. Usually, the titles in the operational

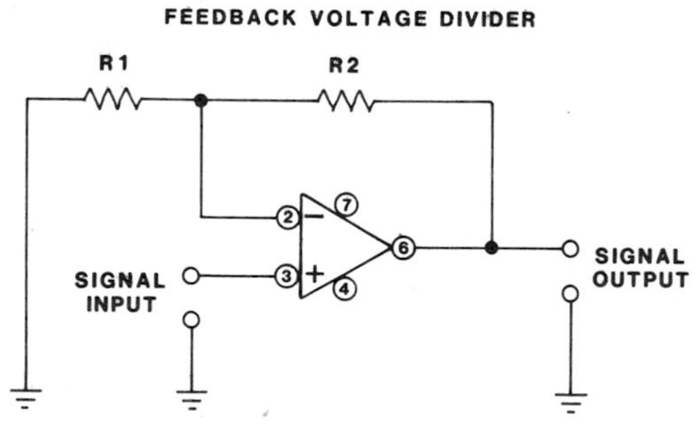

Figure 4–16 OP AMP PRINCIPLES OF OPERATION

amplifier catalogs provide some indication as to the set of parameters that best utilizes the specific skills of an op amp. Some of these are listed next.

1. General purpose (low cost)
2. FET input, low bias current
3. Electrometers (lowest bias current)
4. High accuracy (low drift differential)
5. Chopper amplifiers (lowest drift)
6. Fast wideband
7. High output
8. Isolated op amps
9. Instrumentation (low offset and drift)

The Ideal Amplifier

The search for an ideal amplifier is an exercise in futility. The characteristics of the operational amplifier are good enough, however, to allow us to treat it as ideal. The following are some amplifier properties that make this so. (The reader must realize that these ratings are seldom, if ever, reached.)

1. *Gain*: infinite
2. *Input impedance*: infinite
3. *Output impedance*: zero
4. *Bandwidth*: infinite
5. *Voltage out*: zero (when voltages into each input are equal)
6. *Current entering the amplifier at either terminal*: extremely small

Open-Loop Operation: Op Amp

Consider the *open-loop response* of the amplifier in Figure 4–17. Note that there is no external feedback. Pin 2 is the inverting input and the point at which the signal is input. Pin 3 is the noninverting input.

Pin 3 is grounded. Pin 6 is the signal output. Since the gain is extremely

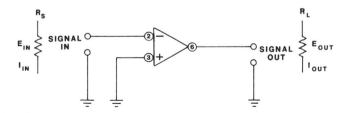

Figure 4-17 OPEN-LOOP OPERATION: OP AMP

high (approaching infinity), the difference in the two inputs must be as near zero as possible. In the ideal amplifier this is the case.

Open-loop gain in some general-purpose operational amplifiers is usually around 10^5. The analysis of open-loop gain is simple. It is the ratio of output to input voltage:

$$\frac{\text{signal out}}{\text{signal in}} = \text{open-loop gain}$$

With no signal, there is no input. Therefore, at pins 2 and 3 the voltage is zero. At pin 6 the voltage is also zero. If pin 2 or 3 takes any signal value, voltage at pin 6 increases to saturation.

A small amount of signal, such as 3 mV, will cause the operational amplifier to traverse to the active region. This is the same effect as in a switching transistor circuit; the output is either high (1) or low (0).

Open-loop operation is used in voltage comparators. Other applications are control-loop-gain scheduling, tracking monitors, and level detection. However, open-loop operation is not nearly as versatile as closed-loop operation.

Input Impedance: Op Amp

The input impedance (Z_{IN}) of an operational amplifier is said to approach infinity. In actuality, each device does have a typical input impedance, such as 1 MΩ and above. Some of the more sophisticated devices may reach an input impedance of as much as 100 MΩ. This value may be determined from the manufacturer's specifications. The input impedance (Z_{IN}) of an op amp is the resistance at either the inverting or the noninverting inputs, with the opposite input at ground.

Output Impedance: Op Amp

The output impedance (Z_{OUT}) of an operational amplifier is said to approach zero. In actuality, each device is different, but for the sake of design this impedance can be assumed to be zero. In specifications, output impedances are said to be 25 to 50 Ω. However, this value can be ignored for most applications.

Because of low output impedance, the output will function as a voltage source and will provide required current for a wide range of loads up to the current limit level (25 mA, typically). With this capability and high input impedance, the operational amplifier becomes an ideal impedance-matching device.

Gain: Op Amp

Probably the most important parameter/specification involved with an amplifier is gain. In fact, that is what the amplifier is all about. An output is derived from some input. The amplifier provides an output that is X amount larger than the amount placed into the input.

Voltage gain with operational amplifiers approaches infinity for the ideal amplifier. For nonideal amplifiers, the voltage gain is adjustable with external resistances. In either event, the following voltage-gain ratio exists:

$$A_v = \frac{V_{OUT}}{V_{IN}}$$

where

$$A_v = \text{voltage gain}$$
$$V = \text{voltage}$$

This may seem rather basic; however, it is essential to most amplifier situations. Two other gain ratios, A_i (current gain) and A_p (power gain), should be stated as fundamental to the amplifier cause:

$$A_i = \frac{I_{OUT}}{I_{IN}}$$

where

$$A_i = \text{current gain}$$
$$I = \text{current}$$

and

$$A_p = \frac{P_{OUT}}{P_{IN}}$$

where

$$A_p = \text{power gain}$$
$$p = \text{power}$$

All three of these equations are simply ratios of one value to another and are dimensionless. All other parameters and specifications evolve around modifications to the gain ratios. The modifications are either desirable or undesirable.

Open-Loop Voltage Gain: Op Amp

Open-loop gain is the voltage gain of the op amp without a feedback but with a load. These values are around 10^5 (not infinity) and are dependent on frequency. The open-loop dc power gain A_p is around 100 decibels (dB), again, not infinite.

Open-loop gain may vary with temperature, voltage supply, and configuration load. Indeed, open-loop gain may vary somewhat with each device within a specified part number. Open-loop gain will always vary with frequency.

Closed-Loop Voltage Gain: Op Amp

Signal gain is closed-loop gain and is a simple ratio for both inverting and noninverting input signals. For inverting stages, the signal gain is

$$A_v = \frac{R_F}{R_M}$$

where

$$A_v = \text{voltage gain}$$

$$R_F = \text{feedback resistance}$$

$$R_M = \text{metering resistance}$$

For noninverting amplifiers, the gain is

$$A_v = \frac{R_M + R_F}{R_M}$$

The Op Amp As An Amplifier

Inverting amplifers using *operational amps* are of two basic designs: dc and ac. Both types of amplifiers provide high voltage gain. They are very versatile and are used in automatic control systems, sound systems, communication systems, and instrumentation. Because of the high voltage gains, the devices are usually protected by a current limiter. Inputs to the amplifiers should typically not exceed supply voltages.

The *dc inverting amplifier* using an op amp consists of the op amp, a feedback resistance R_2, and a metering resistance R_1 (see Figure 4–18). Before proceeding, the author would like to remind the reader of several of the basics of op amps:

1. No current flows into the amplifier.
2. The input summing points are at virtual ground.

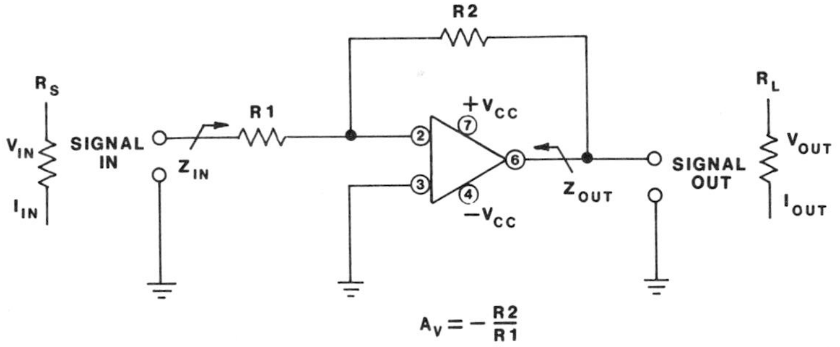

$$A_v = -\frac{R2}{R1}$$

Figure 4–18   INVERTING DC AMPLIFIER: OP AMP

3. The difference voltage between the inputs, inverting and noninverting, is zero.

4. With a feedback attached (closed loop), the negative input signal will be driven to the positive reference input.

To match impedances in a discrete amplifier, the typical situation was to ensure that output impedance of the first stage or source was equal to or less than the second-stage input impedance. Input impedance of an inverting dc amplifier is not infinite but typically around 40 or 50 times the source impedance as determined by R_1. This resistance decision is not absolute. It can vary as long as input impedance is much greater than the source resistance. Output impedances are very low (25 to 50 Ω, typically) and therefore can be ignored in design.

Procedures for design are basically as follows:

1. Calculate the voltage gain required.

$$A_v = \frac{V_{OUT}}{V_{IN}} = \frac{R_2}{R_1}$$

2. Choose R_2, typically at 50 times the source impedance.
3. Calculate feedback resistor $R_2 = A_v R_1$.

The ac version of the inverting amplifier design is the same procedure with a capacitor added as an input coupling device. The choice of a capacitor is dependent on the lowest frequency the circuit may encounter.

Noninverting amplifiers using operational amplifiers provide high voltage gain. Since this is the case, the output is usually protected by a current limiter. Inputs to the noninverting amplifier should typically not exceed the supply voltage. Large-signal voltage gain may approach values such as 100,000. Like inverting amplifiers, noninverting amplifiers, are used in automatic control systems, sound systems, communication systems, and instrumentation. Also like inverting amplifiers, noninverting amplifiers are basically of two types: dc and ac.

The dc noninverting amplifier using an op amp consists of the op amp, a feedback resistance R_2, and a metering resistance R_1. The input impedance at pin 3 is infinite (see Figure 4–19). The noninverting amplifier has the advantage of near-infinity input, the same as the voltage follower. It is therefore used as a buffer for impedance matching. Its output impedance can be said to be zero.

Low-output impedances are generally attributed to a common-collector (emitter-follower) transistor configuration in the output of the amplifier. In Figure 4–19 the R_1 resistor is chosen to match the source impedance and is placed between the inverting input pin 2 and ground. Feedback is arranged between output pin 6 and inverting pin 2. Power, as shown, is applied to pins 4 and 7.

If one of the power supplies becomes disconnected or open, a large voltage

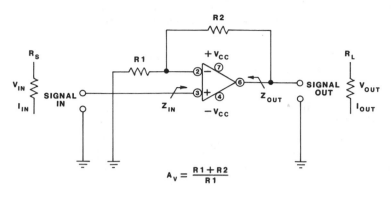

Figure 4-19   NONINVERTING DC AMPLIFIER: OP AMP

may appear at output pin 6. This is dependent on the device used. Voltage gain is

$$A_v = \frac{R_1 + R_2}{R_1}$$

Procedures for design are basically as follows:

1. Calculate the voltage gain required.

$$A_v = \frac{V_{OUT}}{V_{IN}} = \frac{R_1 + R_2}{R_1}$$

2. Choose R_1 to equal the source resistance.
3. Calculate feedback resistor $R_2 = A_v R_1 - R_2$.

The ac version of the noninverting amplifier design is the same procedure with a capacitor added as an input coupling device. The choice of a capacitor is dependent on the lowest frequency the circuit may encounter.

Other Operational Amplifier Circuits and Topics

As you may have determined, the variety of circuits using the operational amplifier is unending. However, a list of some other applications is provided for reference.

1. Integrator
2. Differentiator
3. Astable multivibrator
4. Bistable multivibrator
5. Monostable multivibrator
6. Comparator
7. Norton amplifier
8. Pulse-, triangle-, and shine-wave generators
9. Active filters

10. D/A and A/D conversion   14. Peak detection
11. Voltage regulators   15. Phase-locked loops
12. Log and antilog amplifiers   16. Summing amplifiers
13. Sample-and-hold circuits   17. Multipliers/dividers

INSTRUMENTATION AMPLIFIERS

Transducers are not usually electrically cooperative. The transducer user is plagued by nonlinear devices, electrical interference, high output impedances, inconvenient voltage ranges, and unbalanced outputs. Furthermore, the transducer is subjected to noise and temperature changes, and often works in remote locations. The use of the op amp for a wide variety of applications has been a boon to the industry. Costs of general-purpose devices are very low, with increases only for speed and precision. The general-purpose device is versatile and can do other things than simply amplify. The transducer, however, with its special problems, has created the need for something better. Use of the op amp for all applications may not be the answer.

The practical transducer shown in Figure 4–20 uses differential connections to a bridge. The output impedance of the bridge is not zero and is unbalanced. Long leads add more resistance and noise pickup is hard to avoid. In a differential amplifier such as shown in Figure 4–21, *common-mode rejection* (CMR) is dependent on the ratio matching of the external resistances. The op amp configuration may not be able to reject the common-mode noise. It is for this reason that the instrumentation amplifier was created. The *instrumentation amplifier* has the qualities required for bridge and transducer applications, high input impedance, high gain, and uncommonly high common-mode rejection (CMR). The only trade-off is that the gain of the instrumentation amplifier is held within a specified range. Gain values are usually under 1000. This is really not a problem, however. If better gain, control, and quantity are required, a second amplifier stage may be added at the output.

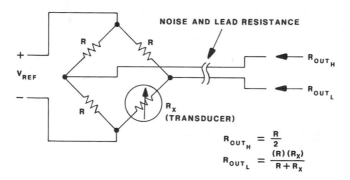

Figure 4–20 PRACTICAL TRANSDUCER CONNECTION TO A BRIDGE

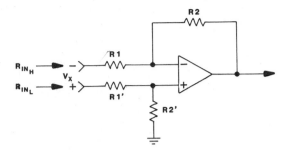

Figure 4–21 DIFFERENTIAL AMPLIFIER OP AMP CONFIGURATION

Characteristics of the Instrumentation Amplifier

An instrumentation amplifier is a high-CMR amplifier with high input impedance. It is committed to a fixed-gain block. The amplifier measures the difference between two voltages placed at input terminals and amplifies it (precisely) (see Figure 4–22).

The output is taken from a pair of external terminals. If the inputs (+ and −) are equal, the output will be zero. Gain is set by the gain resistance R_G and trimmed (in some instrumentation amplifiers) by the gain trim pot R_S. *Offset* may be trimmed if necessary.

The typical instrumentation amplifier has the following characteristics:

1. High input impedance
2. Low bias current
3. High common-mode rejection (CMR)

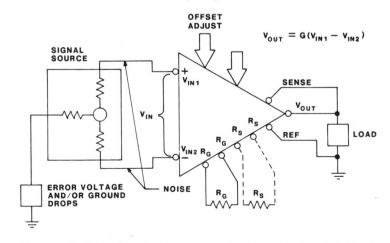

Figure 4–22 BASIC INSTRUMENTATION AMPLIFIER BLOCK DIAGRAM

4. Balanced differential inputs

5. Gain selected by a gain-setting resistor R_G and gain trim (stabilizing resistor) R_S

Gain.   Gains for instrumentation amplifiers are limited within a range of values. For example, the gain range for a typical instrumentation amplifier is from 1 to 1000. Each device is provided with a gain equation. Gain is set with an external gain resistance. The gain formula for the amplifier is as follows:

$$A_v = 1 + \frac{200,000}{R_G} \text{ V/V}$$

where

$$A_v = \text{voltage gain}$$

$$R_G = \text{external gain resistance}$$

The calculation for the gain resistance is as follows:

$$R_G = \frac{200,000}{A_v - 1}$$

Example for gain

$$A_v = 100:$$

$$R_G = \frac{200,000}{100 - 1}$$

$$= 2020 \ \Omega$$

Common-mode rejection.   Common-mode rejection with instrumentation amplifiers is the measure of output voltage caused when both inputs are changed by the same amount. *Common-mode rejection* (CMR) is the log of the ratio CMRR. Common-mode rejection ration (CMRR) is the ratio of signal gain to common-mode gain. In comparison, CMRR of 10,000 is equal to a CMR of 80 dB. Also, a CMRR of 10,000,000 is equal to a CMR of 140 dB. Figure 4–23 shows a simplified CMR. Equal common-mode errors are felt on both inputs. The noninverting input also has a signal present. The instrumention amplifier rejects the common-mode errors and passes only the difference signals.

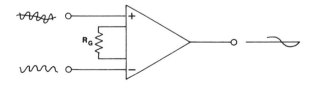

Figure 4–23   COMMON-MODE REJECTION (CMR)

CMR usually increases with high gains. However, at higher gains, bandwidth decreases and CMR becomes dependent on frequency.

Input bias and offset currents.   Input bias currents are used to bias the input transistor stages of the instrumentation amplifier. FET input devices have lower bias currents. Bias currents increase with temperature by as much as double every 10°C. The difference between the input bias currents is the offset current. This difference is the contribution to the offset error.

Voltage offset.   Voltage offset in instrumentation amplifiers is measured at the output. Offset voltage can be adjusted to zero. Offset voltage is equal to a constant plus a term proportional to gain. The constant is the offset at unity gain.

Instrumentation Amplifier Configuration

The basic circuit used in instrumentation is the bridge. In Figure 4–24, the basic bridge is used along with an instrumentation amplifier. You must remember, the instrumentation amplifier has a committed gain block. Gain of the amplifier is set by the gain resistor R_G in this configuration. The gain resistor does not have any circuit connections to the inputs.

This device uses a single gain resistor. Some devices use a gain resistor R_G and a gain stabilizing resistor R_S that provides fine adjustment. In some devices, these resistors are both external. Some devices have only one external resistance. Still others contain all the gain resistors within and use jumpers, switches, or digital logic to perform external programming of gain.

In Figure 4–24, when X is zero, the output should be zero. Any error at the output when X is zero is called *common-mode voltage error*. The amplifier should be able to minimize this by its common-mode rejection (CMR).

The amplifier has a balanced differential input. This means that the output

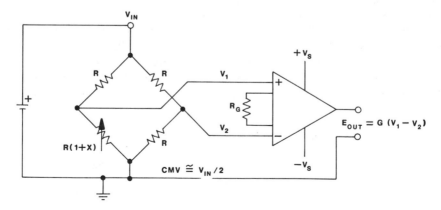

Figure 4–24   BRIDGE INPUT TO AN INSTRUMENT AMPLIFIER

will be proportional to the difference between the two input voltages. The input terminals present a high impedance to the input source. Since the amplifier is a differential amplifier, it can be used to amplify transducer output in bridge networks whether the bridge is balanced or unbalanced. Indeed, it can be the heart of every instrumentation signal conditioner.

Grounding

Instrumentation amplifiers all have differential inputs. There are, however, no *ground paths* for bias current. These must be provided or the bias current may charge with stray capacitance and in turn cause the amplifier output to saturate or to drift. When inputs such as transformers, transducers, or simply ac coupled sources are being amplified, a ground must be supplied. In Figure 4–24 grounds are placed on the input devices and made common with load ground (power supply).

5

Components
of Digital
Signal Conditioning

In our need to control a process using a microprocessor we run into major stumbling blocks. In the world of transducers we are introduced to the dynamic variable, which represents some phenomenon that we would like to measure. That phenomenon is analog and varies with time and circumstances. Microprocessors can only function in this world of analog if there is some method of converting the analog to digital, and vice versa. You see, the digital world has only two states to concern itself with, high and low. The microprocessor does not care to speak to the transducers except in one language, digital. It does not want the responsibilities of noise, variation of signals, and other influences.

It is with these thoughts in mind that *digital signal conditioning* must be considered. Digital signal conditioning is the last stage of the conditioning process. Some texts add conversion of analog to digital (A/D) as just another stage to the process of signal conditioning. In this chapter we treat digital signal conditioning components in an overall view. Those units covered will provide a basic view of digital fundamentals as a link between conditioned analog signals and the data bus.

INTRODUCTION TO DIGITAL

Computer Jargon

The work done by digital electronics is accomplished with or in the computer. It thus makes sense that a person involved with the software of the computer should understand some of the jargon related to digital electronics.

Bit.   Binary numbers are 0 and 1. These numbers are often called bits. The word bit comes from two words, BInary digiT. A number such as 27 is a two-digit number in the decimal system. That same number in the binary (base 2) system is 11011. The base 2 number is a 5-bit number.

External memory.   Memory that is added to the basic computer and interconnected to the CPU is external memory.

Firmware.   These are programs that are fixed within the computer. These programs are called read-only memory (ROM).

Hardware.   The physical parts of the computer are called hardware. These parts may be major components such as the CPU, memory, I/O, and keyboard. Other hardware may be the entire package or wiring. Connectors and printed circuit boards also fall into this category.

Least significant bit (LSB).   The binary digit that has the lowest value is called the LSB (e.g., 11011).

Most significant bit (MSB).   The binary digit that has the most or highest value is called the MSB (e.g., 11011).

Nonvolatile.   If memory is retained when power is turned off, the memory is termed nonvolatile.

Nybble.   A nybble is 4 bits long (e.g., 1101).

Programming.   Instructions input to the computer to carry out a sequence of operations are called programs. These are applied to keyboard, tape, punched cards, and other methods. Programming may be directed to the processing unit or the memory.

RAM.   Random-access memory is that memory which may be read out of, written into, erased, and reused. In other words, it may be modified by almost any means.

ROM.   Read-only memory (ROM) is permanent programmed memory. It cannot be erased or modified. It may be called up (read out) but not written into. ROMs are not affected by shutting off power to the computer.

Software.   Instructions to the computer are called software. The instruction may be in the form of keyboard inputs to memory or processor, magnetic tape, punched cards, paper tape, and so on. Programmed instructions are loaded into the computer memory called random-access memory (RAM).

Volatile.   The word "volatile" has to do with loss of memory. If memory is lost when power is turned off, the memory is termed volatile.

Word.   A word is one or more bits that are related to each other. Words are usually 8, 12, 16, 24, and 64 bits long. The length of the word depends on the computer size. Computer memories are discussed in terms of the number of words they can handle. As an example, an 8K (8000) memory means that the computer will retain approximately 8000 words. These words could be of any length, such as 8 or 16 bits long.

Computer Numbering Systems

The operational states of digital electronics are off or on. This is basically why the relationship between the binary numbering system and digital electronics is so compatible. The *binary numbering system* is based on two digits, 0 and 1. We in the United States and most of the rest of the world depend on the binary system to serve as a vehicle for computer technology. Other numbering systems are used in computer systems, the *hexadecimal system* and *octal system* being two of the most common. The binary numbering system (like any other technical data) is easy to learn once the basics are understood. Let's look at some of the basics and attempt to unravel any mysteries.

Binary number system.   A decimal number is one such as 258. We count inclusively the numbers 0 through 9, which are digits, then move to 10, which means that we have now completed one count through the available digits. We then count through these numbers again to 19, then move to 20, which means that we have completed a second count through the available digits. And so on.

To analyze the structure of a decimal number is rather simple. Let's consider the decimal number 258. This number is made up of 8 units, 5 tens, and 2 hundreds.

$$8 \times 1 = 8$$

$$5 \times 10 = 50$$

$$2 \times 100 = 200$$

$$\text{total} = 258$$

The number 258 can also be analyzed using powers of 10.

$$8 \times 10^0 = 8$$

$$5 \times 10^1 = 50$$

$$2 \times 10^2 = 200$$

$$\text{total} = 258$$

The binary numbering system has two digits, 0 and 1. Binary numbers are expressed in terms of these base 2 numbers. If we were to count in binary, the first count would be from 0 to 1, therefore, 1. We then would be at the end of the available digits and would move to 10, which is the second count. The third count is 11 and we would have again reached the end of the available digits and move to 100. If we continue to count, we very soon have a number such as 10,000, which is actually only 16 counts. Table 5–1 provides us with several of the conversions from digital to decimal for comparison purposes.

TABLE 5–1 Decimal Conversions

Decimal Number	Binary Number
0	0
1	1
2	10
3	11
4	100
5	101
6	110
7	111
8	1000
16	10000
32	100000
64	1000000
128	10000000

The method used to analyze binary (base 2) numbers using powers of 10 is very similar in function to that of decimal numbers. Let's consider the base 2 number 11011. Each position from the right digit to the left digit is represented by a base 2 exponent as follows:

Base 2 Number	Base 2 Exponents		Base 10 Value
1	1×2^0	=	1
1	1×2^1	=	2
0	0×2^2	=	0
1	1×2^3	=	8
1	1×2^4	=	16
		total	27

Therefore, 11011 base 2 = 27 base 10.

Binary-coded decimal system.   Most digital numbers are extremely large in comparison to their decimal equivalent. For example, a digital number such as 11011011 is equal to decimal number 219. The unwieldy digital number is difficult to read and more difficult to convert. Computer software personnel have found that the binary-coded decimal (BCD) system is much easier to cope with. This system uses four binary numbers to represent each decimal digit. For example, the decimal number 219 is represented by binary numbers as follows:

$$0010 \qquad 0001 \qquad 1001$$

$$2 \qquad\qquad 1 \qquad\qquad 9$$

Binary-coded hexadecimal system.   The hexadecimal system is base 16. It is used in digital computers because it is easy to work with and because it can be closely related to the binary system. The first 10 digits of the hexadecimal system are 0 through 9, which is the same as in the decimal system. The last six digits in hexadecimal are the letters A through F. Table 5–2 provides a list of corresponding binary and hexadecimal numbers. These are available on periphery equipment of computers. The hexadecimal system is related to the binary system using four digits. Table 5–3 is a structural analysis of the hexadecimal system.

Binary-coded octal system.   The binary-coded octal (BCO) system is associated with the binary numbering system. The octal system itself is not as satisfactory as the BCO because the octal has only eight digits compared to the decimal system's 10 digits. Table 5–4 provides a list of binary-to-octal equivalents. Table 5–5 is a structural analysis of the binary-coded octal (BCO) system.

TABLE 5–2   Binary-Coded Hexadecimal System

Binary Number	Hexadecimal Number
0000	0
0001	1
0010	2
0011	3
0100	4
0101	5
0110	6
0111	7
1000	8
1001	9
1010	A
1011	B
1100	C
1101	D
1110	E
1111	F

TABLE 5–3 Hexadecimal Structure

Binary position	3	2	1	0
Exponent	16^3	16^2	16^1	16^0
Decimal value	4096	256	16	1

Conversion examples.

Hexadecimal
number

$$138 \quad = 8 \times 16^0 + 3 \times 16^1 + 1 \times 16^2 = 8 + 48 + 256 = 312_{10}$$
$$283 \quad = 3 \times 16^0 + 8 \times 16^1 + 2 \times 16^2 = 3 + 128 + 512 = 643_{10}$$
$$A29 \quad = 9 \times 16^0 + 2 \times 16^1 + A(10) \times 16^2 = 9 + 32 + 2560 = 2601_{10}$$
$$2CB3 \quad = 3 \times 16^0 + B(11) \times 16^1 + C(12) \times 16^2 + 2 \times 16^3 =$$
$$3 + 176 + 3072 + 8192 = 11{,}443_{10}$$

TABLE 5–4 Binary-Coded Octal Systems

Binary Number	Octal Number
000	0
001	1
010	2
011	3
100	4
101	5
110	6
111	7

TABLE 5–5 Binary-Coded Octal Structure

Binary position	3	2	1	0
Exponent	8^3	8^2	8^1	8^0
Decimal value	512	64	8	1

Conversion examples

Octal
number

$$12 \quad = 2 \times 8^0 + 1 \times 8^1 = 2 \times 8 = 16_{10}$$
$$258 \quad = 8 \times 8^0 + 5 \times 8^1 + 8 \times 8^2 = 8 + 40 + 512 = 560_{10}$$
$$3456 \quad = 6 \times 8^0 + 5 \times 8^1 + 4 \times 8^2 + 3 \times 8^3 = 6 + 40 + 256 + 1536 = 1838_{10}$$

Logic Gates

There are five basic logic gates: NOT, AND, NAND, NOR, and OR (see Figure 5–1). The logic gate is the reasoning or decision component of digital electronics. The logic gate simply makes a decision. It decides between functions such as yes or no, true or false. Each input to a gate is considered high or low, HI or LO, 1 or 0. Each output has these same two states. These logic functions are called *binary variables*.

Each logic gate has a mathematical expression from Boolean algebra that describes its function.

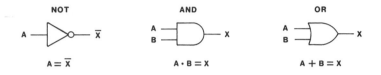

Figure 5–1   LOGIC GATES

NOT gate.   The NOT gate is the simplest of the logic gates. Its mathematical expression is thus: If A is high, then X is not high.

$$A = \bar{X}$$

The input A is either high or low (1 or 0). The output $\bar{X}$ is inverted. The mathematical symbol for this inversion is the bar over the symbol $\bar{X}$.

AND gate.   The AND gate is called the logical product. Its mathematical expression is thus: If A and B are high, then X is high.

$$A \bullet B = X$$

Its mathematical symbol is $\bullet$. The inputs A and B can be either high or low (1 or 0). Only if both inputs are high (1) can there be a low output. The $\bullet$ is often left out: for example, $AB = \bar{X}$.

OR gate.   The OR gate is called the logical sum. Its mathematical expression is thus: If A or B is high, then X is high.

$$A + B = X$$

Its mathematical symbol is +. The inputs A or B can be either high or low (1 or 0). If either is high (1), the output is high (1).

NAND gate.   The NAND gate is the inverse or negation of the AND gate. Its mathematical expression is thus: If A and B are high, then X is not high (low).

$$A + B = \bar{X}$$

Its mathematical symbol is •. The inputs A and B can be either high or low (1 or 0). Only if both inputs are high (1) can there be a low output. The • is often left out: for example, $AB = \bar{X}$.

NOR gate.   The NOR gate is the inverse or negation of the OR gate. Its mathematical expression is thus: If A or B is high, then X is not high (low).

$$A + B = \bar{x}$$

Its mathematical expression is +. The inputs A and B can either be high or low (1 or 0). If either input is high (1), then X is low (0).

Truth Tables

The truth table provides a list of relationships between possible inputs and their outputs. Inputs are either 1 or 0 (high or low). Outputs are either 1 or 0. Inputs are in binary order. Each of the basic gates has fixed truth tables. These tables are given in Table 5–6.

To find the number of binary input combinations for gates with more than one input, powers of 10 are used. For a two-input gate (A, B), there are four binary input combinations, that is, two inputs with two possibilities (1, 0), or 2^2. For a three-input gate (A, B, C), there are eight binary input combinations, that is, three inputs with two possibilities (1, 0), or 2^3. Some of these combinations are listed in Table 5–7.

TABLE 5–6   Basic Gate Truth Tables

NOT GATE		AND GATE			OR GATE			NAND GATE			NOR GATE		
A	X	A	B	X	A	B	X	A	B	X	A	B	X
0	1	0	0	0	0	0	0	0	0	1	0	0	1
1	0	0	1	0	0	1	1	0	1	1	0	1	0
		1	0	0	1	0	1	1	0	1	1	0	0
		1	1	1	1	1	1	1	1	0	1	1	0

TABLE 5–7   Binary Input Combinations

Number of Inputs	Power of 10	Binary Combinations
1	2^1	2
2	2^2	4
3	2^3	8
4	2^4	16
5	2^5	32
6	2^6	64
7	2^7	128
8	2^8	256
9	2^9	512
10	2^{10}	1024

ANALOG-TO-DIGITAL CONVERTERS

The *A/D converter* is used to interface the signal-conditioned analog signal to the data bus, then, in turn, to the computer or microprocessor. The necessity of this operation is to ensure that the analog signals from the transducer are proportional to the digital (binary) words representing these analog signals. We shall discuss some of the methods of conversion beginning with simple comparator.

Comparators

The comparator is the simplest form of A/D converter. The comparator is an integrated-circuit operational amplifier that operates in open-loop mode. The output of the comparator saturates positive if the noninverting input is greater than the inverting input. Similarly, the output will saturate negative if the noninverting input is less than the inverting input. If the output is to be used for logic purposes, it may be clamped to some dc level.

In Figure 5–2A the simplest form of the comparator is illustrated. The inverting input is referenced to some dc level by V_{REF}. The input analog signal is connected to the noninverting input as V_{IN}. When the analog signal V_{IN} exceeds the value of V_{REF}, the output V_{OUT} saturates at a value set by the clamping voltage across R_C. In the figure, this happens to be 5 V. The output in logic also represents a logic 1 or high. If the analog input is smaller than reference V_{REF}, the output would be at 0 V, which represents a logic 0 or low.

In Figure 5–2B a second and more versatile comparator is shown. On this configuration, the reference V_{REF} has negative and positive supply voltages. Here the reference may be set from some negative to some positive level. With this in mind, the input must reach a more negative or positive level to saturate the output. The output is high (1) or low (0) depending on the size of the input and its polarity.

In Figure 5–2C the noninverting input is connected to a positive power supply and a voltage divider R_1 and R_2. The voltage-divider resistances are chosen

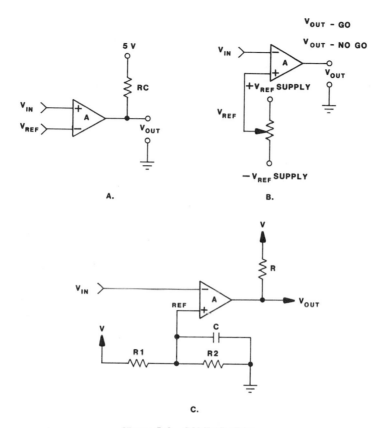

Figure 5–2   COMPARATOR

to set the desired threshold level. The capacitor C is in parallel with R_2. The purpose of the capacitor is to prevent ringing. *Ringing* is a decaying oscillation that may occur at transition when the comparator switches from high to low, and vice versa. As in Figure 5–2A, the input increases below and above the reference voltage to provide a low (0) and a high (1) output.

Logically speaking, the comparator provides us with a 1-bit converter. It simply tells us whether an analog value is above or below some set point or threshold. With this truism we can state the following:

$$V_{IN} > V_{REF} = \text{logic } 1$$

$$V_{IN} < V_{REF} = \text{logic } 0$$

The Comparator Used In a Simple Alarm System

The comparator is used in simple logic systems to perform functions such as interfacing with a logic gate to set off an alarm. In Figure 5–3 a vat of liquid is being monitored for temperature and liquid level.

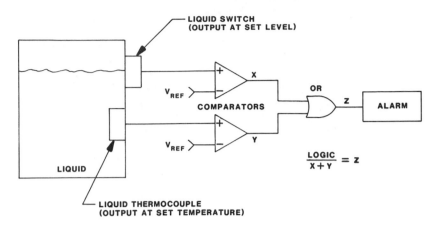

Figure 5-3   COMPARATOR IN A SIMPLE ALARM SYSTEM

If the temperature of the liquid rises above a threshold level set by V_{REF}, the comparators output will saturate high (1). This action will, in turn, cause the OR gate output to go high (1) and set the alarm. Furthermore, if the liquid level increases beyond the threshold set by V_{REF}, its comparator output will saturate high (1). This action will also cause the OR gate output to go high (1) and set the alarm.

This simple comparator is a 1-bit conversion method for a pair of variables. In the first condition, the comparator has two logic states:

High (1) when the temperature is above the set point

Low (0) when the temperature is below the set point

In the second condition, the comparator also has two logic states:

High (1) when the liquid level is above the set point

Low (0) when the liquid level is below the set point

An English-language statement can thus be written from these conditions: If the liquid level is high (1) or the temperature is high (1), the alarm will sound. This allows us to create the Boolean equation

$$X + Y = Z$$

Analog-to-Digital Conversion Types

There are two generally accepted types of A/D conversion: the direct type and the indirect type. The *direct conversion* type utilizes a digital-to-analog (D/A) converter as a reference. The analog signal is compared repeatedly and frequently with the output of the D/A converter. The D/A converter's output is changed to accommodate the analog signal. When a comparison corresponds with the analog input, the input has the necessary binary code.

Indirect conversion types transform the analog input into a time or frequency domain. Then digital logic is used to convert the time or frequency into a digital output. There are many approaches of A/D converters used in process control systems. We shall cover just a few of these. Most A/D converters are available in IC form. Discussions here are to be used for understanding their purposes.

Simplified Parallel Analog-to-Digital Converters

The simplest but most expensive of the A/D converters is the parallel A/D. This converter is also called a *simultaneous A/D converter*. Analog signals are applied to several comparators at the same time. The comparators are referenced to a voltage that sets them one least significant bit (LSB) apart (2^0, 2^1, 2^2, etc.). An input voltage will change the state of a particular comparator if that comparator's reference voltage is less than the input. All comparators whose reference voltage is larger than the voltage-divided reference do not change states but remain low. Resistors in the voltage divider set the voltage reference so that the voltages at input reference are equivalent between comparators and additive from one comparator to the other.

Figure 5–4 is a simplified two-bit converter. The figure has three comparators fed by one voltage input. An inverter, and AND gate, and an OR gate provide decoding logic. Resistors provide a voltage divider from 6 V reference to ground. The resistances are chosen so that R has a 1-V drop and $2R$ has a 2-V drop. Reference voltages into the comparator are as follows:

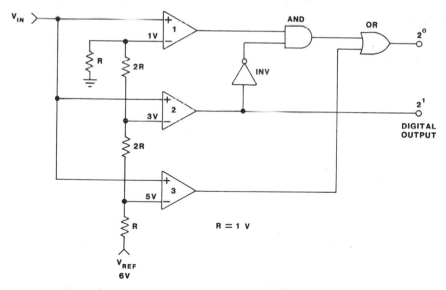

Figure 5–4   SIMPLIFIED PARALLEL 2-BIT CONVERTER

$$\text{Comparator } 1 = 1 \text{ V}$$

$$2 = 3 \text{ V}$$

$$3 = 5 \text{ V}$$

There are two digital outputs, 2^0 and 2^1, hence a 2-bit converter.

With zero signal applied, all comparators will have an output that is high. The inverter, AND gate, and OR gate will have low (0) outputs and the digital word will be 00.

With a 1.5-V signal, the comparator 1 output will go high while the other two comparator outputs will remain low. The inverter output will remain high. The AND gate output will go high and the OR gate output will go high. The digital word will be 01. The following table lists several selected analog voltages and conversion to the digital word.

V_{IN} (V)	Digital Word
0.0	00
1.5	01
3.5	10
5.5	11

To perform conversion of a larger digital word, comparators are added at the rate of $2^n - 1$. For example, an 8-bit word requires $2^8 - 1$ or $256 - 1$ or 255 comparators and associated encoding logic. As you can see, this would become an expensive set of circuits. This entire circuit set is available in single analog-to-digital (A/D) converter chips through standard catalogs.

Successive-Approximation Parallel Feedback A/D Converters

The *successive-approximation* converter consists of a comparator, a clock, a successive-approximation register, a D/A converter, and necessary control logic

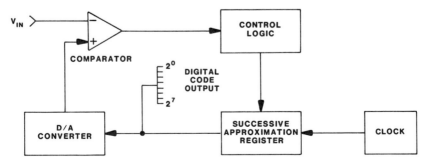

Figure 5–5   SUCCESSIVE-APPROXIMATION PARALLEL FEEDBACK D/A CONVERTER

(see Figure 5–5). The successive-approximation register contains a storage register and a shift register.

This converter compares the input voltage V_{IN} with a set of weighted reference voltage which are the output of the D/A converter. Control logic provides feedback inputs to the successive-approximation register. The register rapidly and repeatedly compares the analog input and the D/A converter input and evaluates the difference. The register outputs this difference in increments of $\frac{1}{2}$, $\frac{1}{4}$, $\frac{1}{8}$, and so on, until the sum of the D/A converter is within one LSB of the actual value of the analog input.

Dual-Slope A/D Integrating Converters

The dual-slope *A/D integrating converter* is highly accurate, rejects noise, and is low cost. It is, however, very slow and can only be used in data acquisition, where speed is not a real factor. The dual-slope A/D converter consists of an integrator, a comparator, a digital counter, and necessary control logic (see Figure 5–6).

The operation of the converter begins with the control logic setting the counter to zero. The electronic switch couples V_{IN} into the integrator and, in turn, to the comparator and counter. The integrator tends to swing negative at a rate that is proportional to V_{IN}. When the output range of the integrator crosses the threshold of the comparator, the counter begins to count. It counts for a fixed period until it overflows. For a constant value of analog input, the slope of the integrator output is proportional to the analog input. At the end of that time period, the

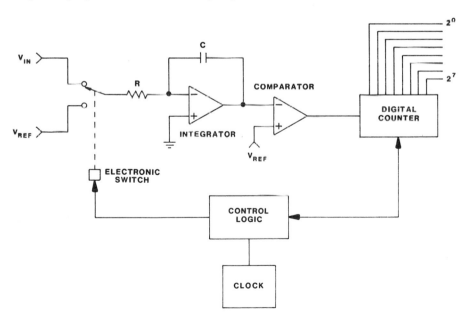

Figure 5–6   DUAL-SLOPE INTEGRATING A/D CONVERTER

integrator is electronically switched to the reference voltage, giving the integrator output ramp a fixed positive slope. The counter counts the time required for the integrator output to reach the comparator's threshold and then stops counting. The value left in the counter is the digital code for the analog input.

Any further detail regarding A/D converters is beyond the scope of this book. Reference is made to the excellent *Analog–Digital Conversion Handbook*, edited by Daniel Sheingold of Analog Devices, Inc. and published by Prentice-Hall.

DIGITAL-TO-ANALOG CONVERTERS

Digital-to-analog (D/A) converters are often required to perform the task of converting an acceptable digital signal into an analog signal for the purpose of driving an analog device. The *D/A converter* accepts a binary word such as 10111001_2 and converts that word into an analog signal. It scales the analog output to zero when all bits of the word are 0s and some maximum value (voltage) when all bits of the word are 1s.

The basic D/A converter system contains an accurate, qualified voltage reference, the D/A converter, and a current-to-voltage converter (see Figure 5–7). The voltage reference supplies the current or voltage. That reference is applied to the D/A converter. The converter is a stack (or ladder) of electronic switches which are controlled by the digital input code and a network of weighted resistances. The switches provide control of the current or voltage derived from the accurate reference. Control produces an output current or voltage that is an analog equivalent of the digital input code. The D/A converter output is fed to the current-to-voltage converter. This converter is usually an operation amplifier or a resistor. The final output is an analog representation of the digital input (see Figure 5–8).

In this figure the output is shown as noncontinuous. Since it is a 3-bit D/A converter, it has eight possible voltage levels. These voltage levels are representative of the digital input code. Each code binary number is a fractional part of the maximum. Each value corresponds to one binary input of 000 to 111, which are

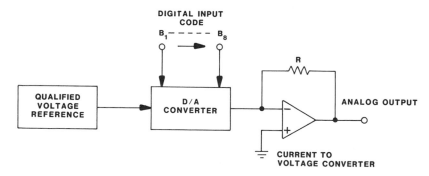

Figure 5–7   BASIC DIGITAL-TO-ANALOG CONVERTER SYSTEM

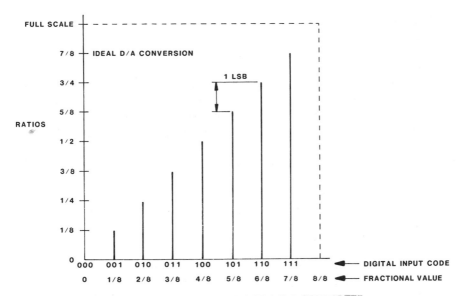

Figure 5–8 ANALOG OUTPUT OF A D/A CONVERTER

one of the input codes. As one can see in the illustration, the full-scale deflection would be some value of voltage, say 10 V. It should be obvious that the more bits in the words applied, the smaller change in fractional parts. Therefore, the greater the number of bits, the greater the resolution.

Weighted Resistor D/A Converters

One of the most useful of the D/A converters is the *weighted resistor converter* (see Figure 5–9). You may recognize the circuit as an op amp summer circuit with the reference input. Inputs to the circuit are binary and in the form of switches S_1 through S_3, hence a 3-bit D/A converter. The MSB is switch S_1. The LSB is switch S_3. The output of the circuit is a summation of current through the three legs. The op amp is created as a buffer and inverts the output.

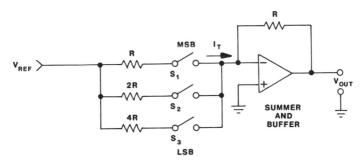

Figure 5–9 THREE-BIT WEIGHTED RESISTOR D/A CONVERTER

$$V_{OUT} = I_T R$$

where I_T is the total current.

Switches in the open position represent digital 0. Switches in the closed position represent digital 1.

A 3-bit D/A converter requires but three resistors. Therefore, if R were selected as 1 KΩ, then $2R$ would be 2 KΩ and $4R$ would be 4 KΩ. A 5-bit D/A converter would require five resistances with the LSB resistor 16 times R or 16 KΩ. An 8-bit D/A converter would require eight resistances with the LSB resistor 128 times R or 128 KΩ.

As the size of the D/A converter increases to say 12 bits, the resistance range becomes extreme and the size of the resistors increase to a maximum of 2.048 MΩ. With large ranges such as this, temperature and size may become a factor. Therefore, the weighted D/A converter is limited to smaller bit inputs.

Ladder Network D/A Converters

The *ladder network* is a resistive summing network sometimes called the R-$2R$ D/A converter (see Figure 5–10). This network is similar in form to the weighted resistor D/A converter. A reference voltage V_{REF} is applied to four switches S_1 through S_4. Networks of resistors of the same values are chosen which allow for closer selection of resistors and better temperature control. Binary inputs close the electronic switches to allow division. The MSB switch directs the reference through only one $2R$ resistor, S_2 through a $2R$ and R, and so on.

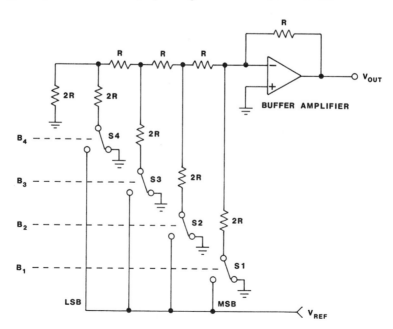

Figure 5–10 LADDER NETWORK (R-$2R$) D/A CONVERTER

This circuit is a 4-bit D/A converter. The digital word 0000 at its lowest voltage level out and 1111 at the highest voltage level. The least significant bit (LSB) is 2^0 and the most significant bit (MSB) is 2^3. The buffer amplifier acts as a summing amplifier where $V_{OUT} = I_T R$.

DATA ACQUISITION SYSTEM

The data acquisition system (DAS) samples dynamic variables from many sources and programs them into the computer. Although the DAS is not signal conditioning as such, we discuss it briefly in this chapter.

The simplified DAS consists of an address decoder, an analog multiplexer, an amplifier, and an A/D converter, together with connecting lines. Its purpose is to interface the conditioned signals into the computer. The process of data acquisition begins with the computer. The computer takes a sample of many different analog sources and evaluates that sample in accordance with preprogrammed instruction. It then outputs that signal to a final control element. The computer selects one of many signal loops. The DAS is the system utilized to perform this task.

Figure 5–11 represents a simplified DAS system. The computer executes a command to fetch contents of a memory. The memory for transducers is an analog input channel. Control lines and address lines are connected to the address decoder. The decoded address is sent to the analog multiplexer and selects the channel connected to that analog input line. The multiplexer is essentially a solid-state switch.

An amplifier, by gain change, compensates for input levels of the analog

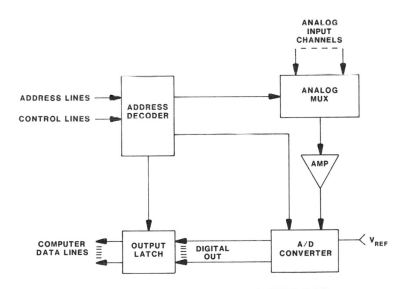

Figure 5–11   DATA ACQUISITION SYSTEM (DAS)

signals, if any. The analog signals must all fall within the range of the A/D converter. The A/D converter accepts the multiplexed and amplified signals and converts them to a compatible digital signal. These signals are latched onto computer data lines by a flip-flop.

DATA OUTPUT MODULE

Just as the data acquisition system (DAS) is used to input data to the computer, the *data output module* (*DOM*) (Figure 5–12) is used to output signals to be used as set point adjustment or for control purposes. Note the similarity in system configuration. The DOM is used in a many-channel system where sampling is necessary.

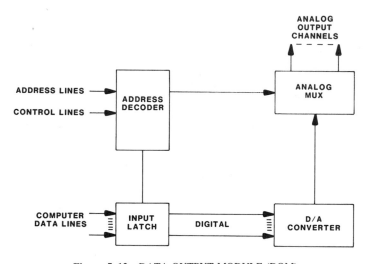

Figure 5–12   DATA OUTPUT MODULE (DOM)

As with the DAS, the DOM begins with the computer. The address decoder is directed to select a particular data line. The computer writes information into a memory location which is latched in and then converted to an analog voltage by the D/A converter. A multiplex switch switches the output to one of many selected channels. The latch controls the data for long enough to ensure conversion of the signal.

SAMPLE-AND-HOLD CIRCUITS

Sample-and-hold (S/H) circuits are used extensively in data acquisition systems. Their major function is to provide a stored analog input to an analog-to-digital (A/D) converter. S/H circuits may also be used to smooth the outputs of a digital-to-analog (D/A) converter which may have been exposed to spikes or during conversions of the D/A from one analog level to another.

The fundamental sample-and-hold (S/H) circuit consists of a pair of electronic switches (S_1 and S_2 in Figure 5–13), a capacitor, and two operational amplifiers in voltage-follower configurations. The signal V_{IN} is applied to the unity-gain operational amplifier stage. The op amp A_1 provides a high impedance for the analog input and a low impedance to the switches and capacitor. The switch S_1 is closed by the computer, allowing the analog signal to charge the capacitor. The circuit is in the sample mode. Sampling continues to charge the capacitor for a specified time, at which the computer opens the switch S_1. During sampling the output may be measured by the data acquisition system through the op amp A_2. This op amp provides a high-impedance input, preventing the capacitor from discharging. Whatever the voltage was across the capacitor when the switch S_1 was opened will remain regardless of the varying input. When the computer decides that it has accomplished its measurement, the switch S_2 is closed long enough to discharge the capacitor to ground, then reopened. The switch S_1 is closed again to continue sampling. When S_1 is closed, sampling takes place. When S_1 is open, the circuit is in hold mode.

The two operational amplifiers in the buffer configuration provide isolation for the S/H switches and capacitor. The output signal tracks the input except for a short period called *acquisition time*, that is, the time for the output to achieve specified level. The output cannot hold the sampling level indefinitely, for the capacitor charge will tend to decay over time.

Input to the sample-and-hold circuit may be a direct analog input, or in the case of a multichannel data acquisition system, may be driven by an analog multiplexer. The multiplexer facilitates the sampling of analog signals from several sources and time shares the inputs into the sample and hold circuits. Then, in turn, into an analog-to-digital (A/D) converter.

The sample-and-hold circuits provide a means for rapidly changing signals to be held for a time period necessary for conversion of analog to digital. However, the S/H circuit adds to software complexity, for it must account for time and produce commands that place the S/H in sample or hold modes.

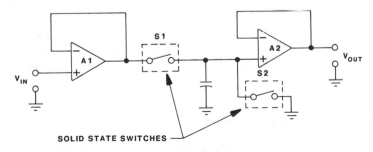

Figure 5–13 SAMPLE-AND-HOLD CIRCUIT

6

Transducer Signal Conditioning

The term *signal conditioning* was defined in some detail in Chapters 4 and 5. The definition of the term is not as precise as you may wish it to be nor as complex as some documentation would lead you to believe. Signal conditioners are used with transducers and may include conversion of the transducer analog signal to a compatible level to be converted to digital and finally to the digital bus. A second definition would have signal conditioning be the entire process of conversion from the transducer signal to the digital bus. For the purposes of this chapter, we consider the latter as being a worthy definition. We shall discuss a variety of transducer types and follow their signal path from the dynamic variable to the digital bus. This chapter acts as an interface between the transducer and the bus. Conditioning diagrams in this chapter include analog-to-digital (A/D) converters and parallel input/output (PIO) circuit blocks which are common to most transducer signal conditioner circuits. The purposes of these common blocks (A/D converters and the PIO) are presented at the beginning of the chapter and will not be repeated.

Within the block diagrams throughout this chapter are small circled numbers. These numbers relate to electronic circuits which are presented in Appendix A. Power supplies are not shown. They may be of any variety as long as they are compatible in power capability.

COMMON BLOCK DESCRIPTION

The common blocks described below are the analog-to-digital (A/D) converter and the parallel input/output (PIO). These two circuits are common to most of the configurations described in this chapter. Repetition cannot be avoided in a reference text. The reader does not need to (nor would want to) jump around in a text more than is absolutely necessary.

Analog-to-Digital Converter

The A/D converter has one basic purpose. The unit converts the signal-conditioned voltage to a digital signal or word that the computer can understand. The A/D converter does this with two operations, by separating the analog signal into parts of equal value and assigning these equal parts a digital word. Separation of the analog signals is called quantization. Assume that the input range to an 8-bit A/D converter is 0 to 10 V. This 8-bit A/D converter will produce a digital word of 00000000_2 or 0 decimal equivalent for a 0-V input. For a 10-V input, the A/D converter will produce a digital word 11111111_2. This digital word represents 256 numbers from 0–255.

The 256 represents the number of fractional parts that quantization will consider. Therefore, the least significant bit (LSB) will be 1/256 of 10 V or 0.0039 V. A midscale input of 5 V would have a 10000000_2 output.

A more detailed explanation of the analog-to-digital (A/D) converter is given in Chapter 5 in the section "Analog-to-Digital Converters". The A/D converter integrated circuits usually accompany the microprocessor chip or are purchased along with the microprocessor chip.

Parallel Input/Output

The purpose of the *PIO* is to interface the A/D converter directly with the computer buses. The stage must provide the "handshaking" between the A/D converter and the computer by controlling data bus traffic direction and signaling the computer when it is ready to release a digital word representing a measured variable.

Generally, the *PIO* is a single IC and is part of the computer's IC chip set. The PIO recognizes when it is to act/react via the address lines from the bus. The PIO performs actions in response to instructions from the control bus. Information is taken or given by way of the data bus. Input to the PIO is a digital word from the A/D converter. Output from the PIO is a digital word to the data bus.

A more detailed explanation of the PIO is found in Chapter 8 in the section "The Microprocessor". PIO integrated circuits usually accompany the microprocessor chip and are therefore described along with the microprocessor.

POTENTIOMETERS

The potentiometric transducer consists of a resistive element and a movable wiper. The measurand (some type of force) operates to move the wiper. The wiper makes contact with the resistance. As the force moves the wiper across the resistance, the relationship of resistance between the wiper and one end of the resistance changes. The change is proportional to position of the wiper. The wiper is attached to a shaft or slider. Since shaft movement is controlled by the measurand (force), the voltage felt at the wiper is proportional to that position. The voltage-divider principle is utilized.

The potentiometer is utilized in various measurement applications, including the following:

1. Linear displacement
2. Linear position
3. Linear velocity
4. Linear acceleration
5. Angular displacement
6. Pressure

Potentiometer Conditioning

The signal derived from a potentiometer is alternating or direct current. In the case of the linear potentiometer transducer, the output signal is dc. The signal should be linear with a total error of less than 0.5% for precision work. Some potentiometers have inherent disadvantages due to age, friction, shock, noise, and wear. Resolution, repeatability, temperature error, and linearity are some of the specifications that are usually considered when choosing a potentiometric transducer. To condition the potentiometric signal to the digital bus, the following circuit functions are required (see Figure 6–1). Refer to the common block descriptions at the beginning of this chapter for circuits flagged with asterisks (*).

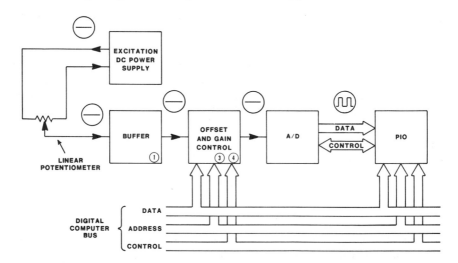

Figure 6–1 ANALOG-TO-DIGITAL CONVERSION: LINEAR POTENTIOMETER

1. Excitation for the transducer and other circuits
2. Isolation of the transducer from other circuitry
3. Amplification
4. Offset control
5. Gain control
6. Scaling control
*7. Analog-to-digital (A/D) conversion
*8. PIO interface with the bus

Buffer.   The buffer serves as a high-resistance input to the signal-conditioning circuit. The buffer prevents loading down of the transducer and thus distortion of the linearity of the transducer. The output of the transducer is fed into the buffer. The buffer isolates the transducer from other circuitry and presents it to the next stage as a voltage proportional to the measurand. Although the input to the buffer may be ac or dc from the transducer, it is still variable (analog).

Offset and gain control.   The offset and gain control circuit has several purposes. One of these purposes is to provide circuitry to eliminate error voltages imposed by the transducer. An offset voltage is applied in opposition to the error to ensure an input signal accuracy with zero error.

A second purpose of this stage is to adjust the amplification of the buffered voltage so that the voltage output is scaled properly. Scaling is required so that the signal is properly proportional to the variable being measured by the transducer.

The gain of the stage, as shown on the block diagram, is controlled automatically by the microprocessor/computer. This method is accomplished with the use of multiplying digital-to-analog converters and latches. A manual method may be provided which provides potentiometers for gain adjustment. The stage can also accommodate additional circuitry to implement a shunt resistor for calibration.

The offset function of this stage can be eliminated by programming the computer to measure the offset and store that value in memory. Each time a measurement is made, the computer subtracts the offset value from the measurand value to achieve the accurate variable.

A more complete implementation of the programming method uses the straight-line formula

$$y = mx + b$$

This allows the computer to calculate the output by storing the offset value b and scale factor m. The measured variable y is then calculated for every buffered voltage x.

A final function of the offset gain control stage is to provide electronic circuitry for filtering noise. Software noise elimination may also be present. Software may be written by performing standard deviation calculations on the buffered voltage inputs.

The input to the offset and gain control is variable (analog). Output from the offset and gain control is directed to the analog-to-digital (A/D) converter. The output is direct current.

Excitation power supply.   The power supply is direct current (dc). Its purpose is to provide excitation for the resistive part of the linear potentiometer. Current flowing through the resistor of the potentiometer allows the wiper to pick off a voltage representing the error or change in measurand. Although not shown on the block diagram, an appropriate power supply is necessary to establish dc power for electronics within the various blocks.

LINEAR VARIABLE DIFFERENTIAL TRANSFORMERS

The linear variable differential transformer (LVDT) is the primary *mutual inductance element*. The LVDT produces an electrical signal that is proportional to the linear displacement of a movable armature or core.

The LVDT has a simple construction. Basically there are two elements involved with the LVDT, the armature and the transformer. The transformer has a stationary coil enclosed in a protective magnetic shield. The armature then moves within the hollow core of the coil.

The coil has a primary winding and two secondaries, wired in series opposition. When the primary is energized by an ac current, the armature (made of a closely controlled magnetic material) induces a voltage from the primary to the secondary windings. The position of the armature within the core of the coil determines the level of the voltage at each secondary. If the armature is placed precisely midway between the two secondaries (null position), the induced voltage in each secondary is equal and opposite, and there is no output. As the armature is moved in either direction away from null, the LVDT produces an output voltage that is proportional to the displacement of the armature from null and whose phase relationship with the primary supply shows whether the armature has moved nearer one end or the other of the coil. Thus, for each position of the armature, there is a definite output voltage, different in level and polarity than for any other position, no matter how slight the difference.

The LVDT is utilized in various measurement applications, including the following:

1. Linear displacement
2. Linear position
3. Linear velocity
4. Linear acceleration
5. Angular acceleration (rotary)
6. Force
7. Torque

8. Vibration displacement
9. Pressure

LVDT Conditioning

The signal derived from a LVDT is alternating current (ac). The signal should be linear with a total error of less than 0.5% for precision work. The LVDT's frequency range may vary from less than 60 Hz to about 20 kHz. Its linear measurement is also broad and can be from 0.00002 to 1000 mm core displacement. The LVDT operates from a null position. As the armature moves from null, its output changes 180°. This provides a position output signal reference. Voltage outputs for various models may vary in range from 25 to 250 mV per millimeter of core motion or more. Linearity of the LVDT is dependent on a reliable power supply. Low-resistance loads critically affect LVDT linearity. A good buffer amplifier may correct linearity problems. Specifications that are necessarily important are linearity, load, winding resistances, frequency range, stroke length versus voltage output, temperature both operating and maximums, repeatability, and hysteresis. Major functions necessary to signal conditioning the LVDT are listed below (see Figure 6–2). Refer to the common block descriptions at the beginning of the chapter for circuits flagged with asterisks (*).

1. Excitation of the transducer and circuits
2. Isolation of the transducer from other circuitry
3. Amplification
4. Demodulation of the ac signal

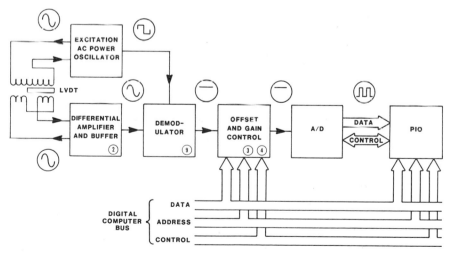

Figure 6–2   ANALOG-TO-DIGITAL CONVERSION: LINEAR VARIABLE DIFFER-
ENTIAL TRANSFORMER (LVDT)

  5. Offset control
  6. Gain control
  7. Scaling control
  *8. Analog-to-digital (A/D) conversion
  *9. PIO interface with the bus

Differential amplifier and buffer.   The differential amplifier provides the proper electronics to measure the voltage across the output of the transducer. The buffer serves as a high-resistance input to the signal-conditioning circuit. It also prevents loading down of the transducer and thereby distortion of its linearity. The buffer isolates the transducer from other circuitry and presents it to the next stage as a voltage proportional to the measurand. The input to the buffer is analog (variable).

Demodulator.   The input to the demodulator is ac from the differential amplifier and buffer. The demodulator provides the circuitry to convert the ac voltage from the buffer into a dc voltage. This action includes the determination of polarity of the dc voltage in relation to the phase of the ac signal input. The demodulator is phase sensitive. It must be synchronized with the power supply excitation.

  Two types of circuits may be used to perform demodulation: the chopper type (full-wave rectification) and sample-and-hold (synchronized peak detector). Either is suitable in this configuration. The demodulator, as a further refinement, removes the inherent quadrature voltages or null voltages of the LVDT. This action aligns the mechanical null (center) position with the electrical null (voltage) output of the transducer.

Offset and gain control.   The offset and gain control circuit has several purposes. One of these purposes is to provide circuitry to eliminate error voltages imposed by the transducer. An offset voltage is applied in opposition to the error to ensure an input signal accuracy with zero error.

  A second purpose of this stage is to adjust the amplification of the buffered voltage so that the voltage output is scaled properly. Scaling is required so that the signal is properly proportional to the variable being measured by the transducer.

  The gain of the stage, as shown on the block diagram, is controlled automatically by the microprocessor/computer. This method is accomplished with the use of multiplying digital-to-analog converters and latches. A manual method may be provided which provides potentiometers for gain adjustment. The stage can also accommodate additional circuitry to implement a shunt resistor for calibration.

  The offset function of this stage can be eliminated by programming the computer to measure the offset and store that value in memory. Each time a measurement is made, the computer subtracts the offset value from the measurand value to achieve the accurate variable.

A more complete implementation of the programming method uses the straight-line formula

$$y = mx + b$$

This allows the computer to calculate the output by storing the offset value b and scale factor m. The measured variable y is then calculated for every buffered voltage x.

A final function of the offset gain control stage is to provide electronic circuitry for filtering noise. Software noise elimination may also be present. Software may be written by performing standard deviation calculations on the buffered voltage inputs.

The input to the offset and gain control is variable (analog). Output from the offset and gain control is directed to the analog-to-digital (A/D) converter. The output is direct current.

Excitation, AC power, and oscillator.   The power supply provides ac power at a specific frequency to excite the primary of the LVDT. Transformer action, you may recall, is part of the function of the LVDT. The power supply must also provide a reference ac and its frequency to the demodulator to synchronize the signal from the LVDT with the demodulator. Although not shown on the block diagram, an appropriate power supply is necessary to establish dc power for electronics within the various blocks.

OPTICAL ENCODERS

Encoders are mechanical-to-electrical transducers whose output is derived by reading a coded pattern on a rotating disk or a moving scale. Encoders are classified into three major categories:

1. The method used to read the code, either contact or noncontact
2. The type of output, either absolute digital word or series of incremental pulses
3. The physical phenomenon employed to produce the output, either electrical conduction, magnetic, capacitive, or optical

The principle of incremental encoder operation is the generation of a symmetric, repeating waveform that can be used to monitor the input motion. The basic components of the optical incremental encoder are the light source, light shutter or chopper, light sensor, signal-conditioning electronics, and shaft bearing assembly. The encoder's mechanical input operates the light shutter, which modulates the intensity of the light at the sensor. The sensor's electrical output is a function of the incident light. The encoder's electrical output is produced from the sensor output by the signal-conditioning electronics and can be either (1) a sine wave, (2) a shaped square wave, or (3) a series of equally spaced pulses produced at regular points on the wave form.

In its most fundamental form the light shutter is an optical slit and a glass plate inscribed with alternating lines and spaces of equal width. When the plate is moved relative to the slit, the light transmitted by the slit rises and falls. The encoder's mechanical input is coupled to the moving plate to operate the shutter. The light source is either a lamp or an LED.

Lack of general information on encoder applications can frustrate the first-time user. Whatever the encoder application, the following basic decisions must be made:

1. What signal level and waveform will be used
2. How the encoder output will be interfaced with the system
3. What type of encoder will be used
4. How the encoder will be mounted
5. How the mechanical input will be coupled to the encoder

The optical encoder is utilized in various applications. Some of these are the following:

1. Linear displacement
2. Linear position
3. Angular displacement
4. Angular velocity

Encoder Conditioning

The encoder transducer must be mounted so that the measurand and the mechanical input to the shaft of the encoder are not overloaded. The shaft may be linear or rotary in construction. The primary objective of coupling is to transmit input motion accurately to the encoder without subjecting its shaft to excessive loads. This coupling may be flexible, gear to gear, rack and pinion, toothed belts and others. The output of the encoder is a digital pulse or word which represents the position of the transducer and its shaft. The number of pulses represent the distance the shaft has moved from its zero position. The bit pattern will provide the user with the shaft position. Specifications include noise, temperature ranges, coupling vibration.

Conditioning for linear displacement applications.

Major functions necessary to signal condition the encoder for linear displacement applications are listed below [see Figure 6–3(A)]. Refer to the common block description at the beginning of this chapter for the circuit flagged with an asterisk (*).

1. Excitation of the transducer and circuits
2. Isolation of the transducer from other circuitry

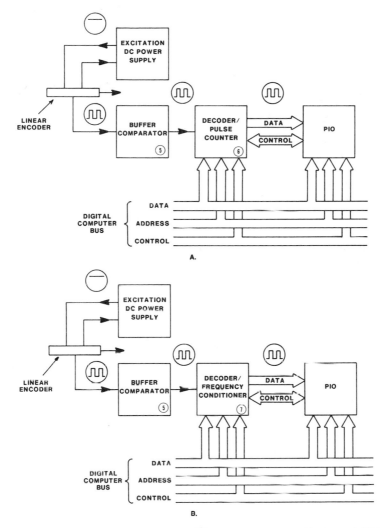

Figure 6-3 DIGITAL-TO-DIGITAL CONVERSION:LINEAR ENCODER: (A) LIN-
EAR DISPLACEMENT (B) LINEAR VELOCITY

3. Amplification
4. Comparator
5. Decoder/pulse counter
*6. PIO interface with the bus

Buffer Comparator.   The buffer serves as a high-resistance input to the
signal conditioning circuit. The buffer prevents loading down of the transducer
and thus distortion of its output. The output of the transducer, an analog voltage,

is fed into the buffer comparator. The buffer isolates the encoder from other circuitry and presents it to the next stage as a voltage proportional to the measurand. The input to the buffer is variable (analog). The comparator is useful to determine whether a signal is above or below threshold. Its output saturates at the highest positive level, providing logic-level outputs. Further, the comparator "cleans up" the leading and falling edges of the transducer's pulse signals.

Decoder/Pulse Counter.   If the transducer has a pulse output, a decoder is used to convert the buffer output into a digital word that is representative of the transducer's position. The other part of this stage is a pulse counter. That circuit counts buffer output pulses. The output of the pulse counter is a digital word representing the number of pulses counted. Since the output of the decoder/pulse counter is digital, there is no requirement for an A/D converter. Therefore, the decoder is connected directly to the PIO.

Conditioning for linear velocity applications.   Major functions necessary to signal condition the encoder for linear velocity applications are listed below [see Figure 6–3(B)]. Refer to the common block description at the beginning of this chapter for the circuit flagged with an asterisk (*).

1. Excitation of the transducer and circuits
2. Isolation of the transducer from other circuitry
3. Amplification
4. Comparator
5. Decoder/frequency conditioner
*6. PIO interface with the bus

Buffer Comparator.   The buffer serves as a high-resistance input to the signal-conditioning circuit. The buffer prevents loading down of the transducer and thus distortion of its linearity. The output of the transducer, an analog voltage, is fed into the buffer comparator. The buffer isolates the encoder from other circuitry and presents it to the next stage as a voltage proportional to the measurand. The input to the buffer is variable (analog). The comparator is useful to determine whether a signal is above or below threshold. Its output saturates at the highest positive level, providing logic-level outputs. Further, the comparator "cleans up" the leading and falling edges of the transducer's pulse signals.

Decoder/Frequency Conditioner.   If the transducer has a pulse output, a decoder is used to convert the buffer output into a digital word that is representative of the transducer's frequency. The other part of this stage is a frequency conditioner. This circuit is used to measure the frequency of output pulses of the buffer. The output of the frequency conditioner is a digital word representing the frequency of the buffer output pulses.

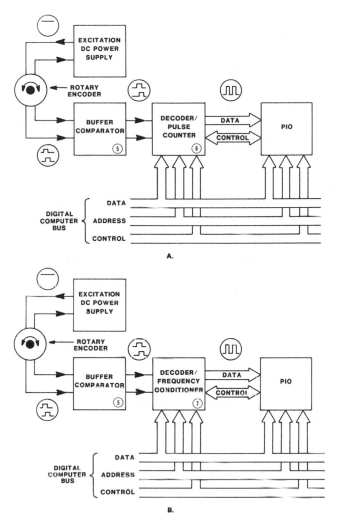

Figure 6–4 DIGITAL-TO-DIGITAL CONVERSION: ROTARY ENCODER: (A) ANGULAR DISPLACEMENT (B) ANGULAR VELOCITY

Conditioning for angular displacement applications. Major functions necessary to signal condition the rotary encoder for angular displacement applications are listed below [see Figure 6–4(A)]. Refer to the common block description at the beginning of this chapter for the circuit flagged with an asterisk (*).

1. Excitation of the transducer and circuits
2. Isolation of the transducer from other circuitry

3. Amplification
4. Comparator
5. Decoder/pulse counter
*6. PIO interface with the bus

Buffer Comparator.   The buffer serves as a high-resistance input to the signal-conditioning circuit. The buffer prevents loading down of the transducer and thus distortion of its linearity. The output of the transducer, an analog voltage, is fed into the buffer comparator. The buffer isolates the encoder from other circuitry and presents it to the next stage as a voltage proportional to the measurand. The input to the buffer is variable (analog). The comparator is useful to determine whether a signal is above or below threshold. Its output saturates at the highest positive level, providing logic-level outputs. Further, the comparator "cleans up" the leading and falling edges of the transducer's pulse signals.

Decoder/Pulse Counter.   If the transducer has a pulse output, a decoder is used to convert the buffer output into a digital word that is representative of the transducer's position. The other part of this stage is a pulse counter. That circuit counts buffer output pulses. The output of the pulse counter is a digital word representing the number of pulses counted. Since the output of the decoder/pulse counter is digital, there is no requirement for an A/D converter. Therefore, the decoder is connected directly to the PIO.

Conditioning for angular velocity applications.   Major functions necessary to signal condition the rotary encoder for angular velocity applications are listed below [see Figure 6–4(B)]. Refer to the common block description at the beginning of this chapter for the circuit flagged with an asterisk(*).

1. Excitation of the transducer and circuits
2. Isolation of the transducer from other circuitry
3. Amplification
4. Comparator
5. Decoder/frequency conditioner
*6. PIO interface with the bus

Buffer Comparator.   The buffer serves as a high-resistance input to the signal conditioning circuit. The buffer prevents loading down of the transducer and thus distortion of its linearity. The output of the transducer, an analog voltage, is fed into the buffer comparator. The buffer isolates the encoder from other circuitry and presents it to the next stage as a voltage proportional to the measurand. The input to the buffer is variable (analog). The comparator is useful to determine whether a signal is above or below threshold. Its output saturates at the highest positive level, providing logic-level outputs. Further, the comparator "cleans up" the leading and falling edges of the transducer's pulse signals.

Decoder/Frequency Conditioner.   If the transducer has a pulse output, a decoder is used to convert the buffer output into a digital word that is representative of the transducer's frequency. The other part of this stage is a frequency conditioner. This circuit is used to measure the frequency of output pulses of the buffer. The output of the frequency conditioner is a digital word representing the frequency of the buffer output pulses.

Excitation DC power supply.   The power supply provides dc power

to the encoder. Although not shown on the block diagram, an appropriate power supply is necessary to establish dc power for electronics within the various blocks.

LINEAR VARIABLE TRANSFORMERS

The linear variable transformer (LVT) consists of a spring-suspended mass (usually a magnet) used as a core within a housing. The housing has a coil of wire surrounding the core. The flux between the core and the coil changes as a result of the motion of the core across the coil field. A shaft connected to the core presents the measurand as a linear motion. The output of the coil is proportional to the velocity of core motion. In some other applications, the coil may be the moving part. The *velocity transducer* produces a self-generating voltage of a magnitude large enough so that little amplification is required. Therefore, being voltage self-generating, no power supply excitation voltage is required. The LVT transducer is generally larger than the engineer would like, and it has the disadvantage of not being able to function at low frequencies (less than 15 Hz or so). Furthermore, the output voltage at large frequencies (more than 1000 Hz) tends to be small and awkward to work with.

The LVT is utilized in various applications. Some of these are listed below.

1. Linear velocity
2. Linear acceleration
3. Angular displacement
4. Vibration velocity

LVT Conditioning

The signal derived from an LVT is dc like and analog in nature. *Velocity* is often related to *acceleration* and displacement and is useful in the study of machine vibration. The transducer has a high-output, low-source impedance and is comparatively inexpensive. Specifications include environmental aspects and

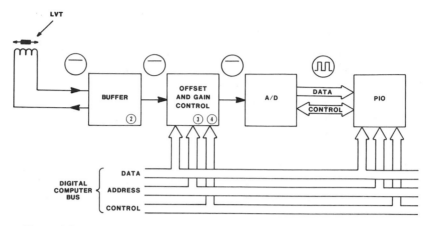

Figure 6–5 ANALOG-TO-DIGITAL CONVERSION: LINEAR VARIABLE TRANS-
FORMER (LVT)

velocity response. Major functions necessary to signal conditioning the LVT are
listed below (see Figure 6–5). Refer to the common block descriptions at the
beginning of this chapter for circuits flagged with asterisks (*).

1. Isolation of the transducer from other circuitry
2. Amplification
3. Offset control
4. Gain control
5. Scaling control
*6. Analog-to-digital (A/D) conversion
*7. PIO interface with the bus

Buffer.   The buffer serves as a high-resistance input to the signal-conditioning circuit. The buffer prevents loading down of the transducer and thus distortion
of the linearity of the transducer. The output of the transducer is fed into the
buffer. The buffer isolates the transducer from other circuitry and presents it to
the next stage as a voltage proportional to the measurand. Although the input to
the buffer may be ac or dc from the transducer, it is still variable (analog).

Offset and gain control.   The offset and gain control circuit has several
purposes. One of these is to provide circuitry to eliminate error voltages imposed
by the transducer. An offset voltage is applied in opposition to the error to ensure
an input signal accuracy with zero error.

A second purpose of this stage is to adjust the amplification of the buffered
voltage so that the voltage output is scaled properly. Scaling is required so that
the signal is properly proportional to the variable being measured by the transducer.

The gain of the stage, as shown on the block diagram, is controlled automatically by the microprocessor/computer. This method is accomplished with the use of multiplying digital-to-analog converters and latches. A manual method may be provided which provides potentiometers for gain adjustment. The stage can also accommodate additional circuitry to implement a shunt resistor for calibration.

The offset function of this stage can be eliminated by programming the computer to measure the offset and store that value in memory. Each time a measurement is made, the computer subtracts the offset value from the measurand value to achieve the accurate variable.

A more complete implementation of programming method uses the straight-line formula

$$y = mx + b$$

This allows the computer to calculate the output by storing the offset value b and scale factor m. The measured variable y is then calculated for every buffered voltage x.

A final function of the offset gain control stage is to provide electronic circuitry for filtering noise. Software noise elimination may also be present. Software may be written by performing standard deviation calculations on the buffered voltage inputs.

The input to the offset and gain control is variable (analog). Output from the offset and gain control is directed to the analog-to-digital (A/D) converter. The output is direct current.

Excitation DC power supply. Note that there is no power supply for excitation of the transducer, as it is self-generating. Although not shown on the block diagram, an appropriate power supply is necessary to establish dc power for electronics within the various blocks.

STRAIN GAGES

The purpose of a strain gage is to detect the amount of length displaced by a force member. The strain gage produces a change in resistance that is proportional to this variation in length. Strain gages are usually installed as a part of a Wheatstone bridge for electrical-circuit applications.

There are two basic strain-gage types: the bonded and the unbonded. The *bonded gage* is entirely attached to the force member by an adhesive of some sort. As the force member stretches in length, the strain gage also lengthens.

The *unbonded gage* has one end of its strain wire attached to the force member and the other end attached to a force collector. As the force member stretches, the strain wire also changes in length. Each motion of length, either with bonded or unbonded gages, causes a change in resistance. Strain gages are made from metal and semiconductor materials.

Strain gages are utilized in various applications, some of which are the following:

1. Linear displacement
2. Linear position
3. Linear acceleration
4. Angular acceleration
5. Force
6. Torque
7. Vibration, torsional
8. Pressure

Strain Gage Conditioning

Strain gages are accurate, may be excited by alternating or direct current, and have excellent static and dynamic response. The signal out of a strain gage is very small, but this disadvantage may be corrected with good periphery equipment. They may be very small and operate in frequency ranges up to 1 kHz. Specifications include shock capabilities, frequency response, and acceleration. Figure 6–6 shows the strain gage transducer in bridge application circuitry. The strain gage in this configuration uses the voltage-divider concept. Major functions necessary to signal condition the strain gage transducer are listed below (see Figure 6–6). Refer to the common block descriptions at the beginning of this chapter for circuits flagged with asterisks (*).

1. Excitation of the transducer and circuits
2. Isolation of the transducer from other circuitry
3. Amplification
4. Offset control
5. Gain control
6. Scaling control
*7. Analog-to-digital (A/D) conversion
*8. PIO interface with the bus

Differential amplifier and buffer. The differential amplifier provides the proper electronics to measure the voltage across the output of the transducer. The buffer serves as a high-resistance input to the signal-conditioning circuit. It also prevents loading down of the transducer and thus distortion of its linearity. The buffer isolates the transducer from other circuitry and presents it to the next stage as a voltage proportional to the measurand. The input to the buffer is analog (variable).

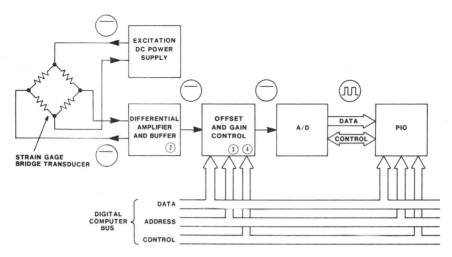

Figure 6–6 ANALOG-TO-DIGITAL CONVERSION: STRAIN GAGE BRIDGE TRANSDUCER

Offset and gain control. The offset and gain control circuit has several purposes. One of these purposes is to provide circuitry to eliminate error voltages imposed by the transducer. An offset voltage is applied in opposition to the error to ensure an input signal accuracy with zero error.

A second purpose of this stage is to adjust the amplification of the buffered voltage so that the voltage output is scaled properly. Scaling is required so that the signal is properly proportional to the variable being measured by the transducer.

The gain of the stage, as shown on the block diagram, is controlled automatically by the microprocessor/computer. This method is accomplished with the use of multiplying digital-to-analog converters and latches. A manual method may be provided which provides potentiometers for gain adjustment. The stage can also accommodate additional circuitry to implement a shunt resistor for calibration.

The offset function of this stage can be eliminated by programming the computer to measure the offset and store that value in memory. Each time a measurement is made, the computer subtracts the offset value from the measurand value to achieve the accurate variable.

A more complete implementation of programming method uses the straight-line formula

$$y = mx + b$$

This allows the computer to calculate the output by storing the offset value b and scale factor m. The measured variable y is then calculated for every buffered voltage x.

A final function of the offset gain control stage is to provide electronic circuitry for filtering noise. Software noise elimination may also be present. Software may be written by performing standard deviation calculations on the buffered voltage inputs.

The input to the offset and gain control is variable (analog). Output from the offset and gain control is directed to the analog-to-digital (A/D) converter. The output is direct current.

Excitation DC power supply.   The dc power supply provides excitation to the bridge circuit. Although not shown on the block diagram, an appropriate power supply is necessary to establish dc power for electronics within the various blocks.

PIEZOELECTRIC TRANSDUCERS

The piezoelectric transducer utilizes a crystalline structure bonded to a mass or a center post. The face of the crystal is attached to the moving element which, in turn, is moved by some dynamic variable. When the moving element has deflection, the mass causes some change across the face of the crystal. The change distorts the crystal, altering its shape and therefore its output frequency. A shear stress on the crystal deforms the crystal in shear. Compression varies the stress put on the crystal by its support structure. The *crystalline structure* is either ceramic or quartz. Both of these self-generating materials produce a large electrical charge. Ceramics are more sensitive than quartz. With a small mass and stiff members (or spring tension) the piezoelectric transducer is highly efficient. Its response is fairly constant for acceleration inputs. If the mass is large and structural members (including spring) are flexible, the mass requires damping.

Piezoelectric transducers are utilized in various applications, including the following:

1. Linear velocity
2. Linear acceleration
3. Force
4. Sound
5. Pressure

Piezoelectric Transducer Conditioning

Signal conditioning for the piezoelectric transducer requires a frequency-to-voltage conversion. The dynamic variable attached to the moving element of which the crystal(s) are attached changes the stress on the crystal, which alters its frequency output. The output is fed directly into a frequency-to-voltage converter circuit. There are no excitation voltage requirements for this transducer, as it is self-generating.

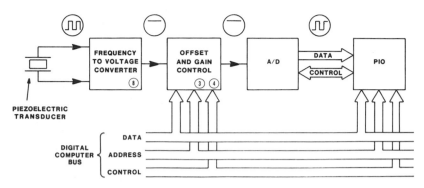

Figure 6-7 ANALOG-TO-DIGITAL CONVERSION: PIEZOELECTRIC

Specifications for the piezoelectric transducer are sensitivity, frequency, range, resonant frequency, amplitude range, shock rating, temperature ranges, and mass. It is important that the user of a piezoelectric transducer be aware of size restrictions and the band of frequencies in which a desired transducer is operational. Other major functions that are required for piezoelectric transducer conditioning are listed below (see Figure 6-7). Refer to the common block descriptions at the beginning of this chapter for circuits flagged with asterisks (*).

1. Isolation of the transducer from other circuitry
2. Amplification
3. Offset control
4. Gain control
5. Scaling control
*6. Analog-to-digital (A/D) conversion
*7. PIO interface with the bus

Frequency-to-voltage converter. The purpose of the frequency-to-voltage converter is to convert the piezoelectric transducer frequency output into a direct-current (dc) voltage. This voltage is proportional to the transducer output. The input impedance of the frequency-to-voltage converter must be matched to the piezoelectric transducer so as to load the transducer properly and not distort its output signal. Impedance is provided by the transducer manufacturer. Conversion is usually done with a charge amplifier. This amplifier consists of a charge converter and a voltage amplifier. The charge converter is an operational amplifier with an integrating feedback. Some low-impedance output transducers come with electronics built into the case. These provide some amplification but low impedance. Their advantage is low noise; their disadvantage is nonadjustable gain.

The frequency-to-voltage converter used with the capacitive transducer is used to convert the frequency of the oscillator into a direct current proportional to that frequency. This requires a proper selection of resistors and capacitors so as to operate in the frequency range of the oscillator.

Offset and gain control.   The offset and gain control circuit has several purposes. One of these purposes is to provide circuitry to eliminate error voltages imposed by the transducer. An offset voltage is applied in opposition to the error to ensure an input signal accuracy with zero error.

A second purpose of this stage is to adjust the amplification of the buffered voltage so that the voltage output is scaled properly. Scaling is required so that the signal is properly proportional to the variable being measured by the transducer.

The gain of the stage, as shown on the block diagram, is controlled automatically by the microprocessor/computer. This method is accomplished with the use of multiplying digital-to-analog converters and latches. A manual method may be provided which provides potentiometers for gain adjustment. The stage can also accommodate additional circuitry to implement a shunt resistor for calibration.

The offset function of this stage can be eliminated by programming the computer to measure the offset and store that value in memory. Each time a measurement is made, the computer subtracts the offset value from the measurand value to achieve the accurate variable.

A more complete implementation of programming method uses the straight-line formula

$$y = mx + b$$

This allows the computer to calculate the output by storing the offset value b and scale factor m. The measured variable y is then calculated for every buffered voltage x.

A final function of the offset gain control stage is to provide electronic circuitry for filtering noise. Software noise elimination may also be present. Software may be written by performing standard deviation calculations on the buffered voltage inputs.

The input to the offset and gain control is variable (analog). Output from the offset and gain control is directed to the analog-to-digital (A/D) converter. The output is direct current.

Excitation DC power supply.   Note that there is no excitation for the piezoelectric transducer as it is self-generating. Although not shown on the block diagram, an appropriate power supply is necessary to establish dc power for electronics within the various blocks.

PIEZOTRANSISTOR TRANSDUCERS

A piezotransistor transducer contains a seismic mass tied to a stylus. The stylus transmits a force. The force causes a stress on one surface of a transistor. This surface of the transistor is part of a *pn* junction diode. The force causes a current change in the diode, which, in turn, causes the transistor to operate. The change in current is proportional to the amount of force. The piezotransistor transducer monitors acceleration of 400g and less.

Although not as versatile because of power supply requirements as other piezo-type devices, the piezotransistor transducer is utilized in various applications, some of which are as follows:

1. Linear acceleration
2. Force
3. Pressure

Piezotransistor Transducer Conditioning

The output from a piezotransistor transducer is direct current, although analog in function. The dynamic variable alters the position of the stylus, causing a current change in the output of the transistor. The output is fed directly into a differential amplifier and buffer. Specifications for the piezotransistor transducer are sensitivity, shock ratings, temperature ranges, size, and weight. Other major functions that are required for piezotransistor transducer conditioning are listed below (see Figure 6–8). Refer to the common block descriptions at the beginning of this chapter for circuits flagged with asterisks (*).

1. Excitation of the transducer and circuits
2. Isolation of the transducer from other circuitry
3. Amplification

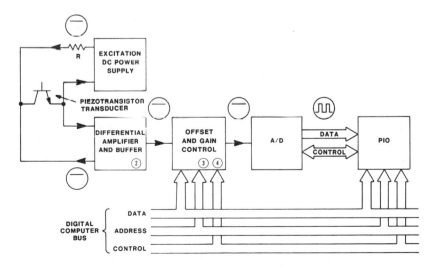

Figure 6–8 ANALOG-TO-DIGITAL CONVERSION: PIEZOTRANSISTOR

4. Offset control
5. Gain control
6. Scaling control
*7. Analog-to-digital (A/D) conversion
*8. PIO interface with the bus

Differential amplifier and buffer.   The differential amplifier provides
the proper electronics to measure the voltage across the output of the transducer.
The buffer serves as a high-resistance input to the signal-conditioning circuit. It
also prevents loading down of the transducer and thus distortion of its linearity.
The buffer isolates the transducer from other circuitry and presents it to the next
stage as a voltage proportional to the measurand. The input to the buffer is analog
(variable).

Offset and gain control.   The offset and gain control circuit has several
purposes. One of these purposes is to provide circuitry to eliminate error voltages
imposed by the transducer. An offset voltage is applied in opposition to the error
to ensure an input signal accuracy with zero error.
  A second purpose of this stage is to adjust the amplification of the buffered
voltage so that the voltage output is scaled properly. Scaling is required so that
the signal is properly proportional to the variable being measured by the transducer.
  The gain of the stage, as shown on the block diagram, is controlled automatically by the microprocessor/computer. This method is accomplished with the use
of multiplying digital-to-analog converters and latches. A manual method may be
provided which provides potentiometers for gain adjustment. The stage can also
accommodate additional circuitry to implement a shunt resistor for calibration.
  The offset function of this stage can be eliminated by programming the
computer to measure the offset and store that value in memory. Each time a
measurement is made, the computer subtracts the offset value from the measurand
value to achieve the accurate variable.
  A more complete implementation of programming method uses the straight-
line formula

$$y = mx + b$$

This allows the computer to calculate the output by storing the offset value b and
scale factor m. The measured variable y is then calculated for every buffered
voltage x.
  A final function of the offset gain control stage is to provide electronic
circuitry for filtering noise. Software noise elimination may also be present. Software
may be written by performing standard deviation calculations on the buffered
voltage inputs.
  The input to the offset and gain control is variable (analog). Output from
the offset and gain control is directed to the analog-to-digital (A/D) converter.
The output is direct current.

Excitation DC power supply. The dc power supply provides excitation for the transducer and other circuits. A special resistor (R) is a current limiter. It is used to protect the transducer from overcurrent. The resistor (R) also sets the operating range of the transducer to achieve optimum power. Although not shown on the block diagram, an appropriate power supply is necessary to establish dc power for electronics within the various blocks.

TACHOMETERS

Angular velocity devices use the tachometer for sensing. In the most basic tachometer, an alternating- or direct-current generator is used as a velocity transducer. The generators produce an *angular velocity* signal from rotary motion. The output signal of a dc generator is a voltage that is directly proportional to the angular velocity of its shaft. The output signal of an ac generator is also a voltage. However, the ac frequency is directly proportional to the angular velocity of the shaft.

In another form, a *magnetic sensor* produces an output frequency, usually from an actuating gear, in direct proportion to rotational speed. The signal generated by the sensor in this mode is completely error-free and can be calculated for any given speed by the formula

$$\text{frequency } (f) = \frac{\text{number of gear teeth} \times \text{rpm}}{60}$$

The frequency thus generated can be converted directly to rpm by means of a frequency counter or digital tachometer. Another method of speed measurement is to change the frequency into a proportional dc current.

Other tachometer types use rotating switches, stroboscopes, strain gages, vibrating reeds, eddy current, and even hydraulics. In all cases the angular velocity is achieved by denoting the direction of rotation and the angular speed of rotation. The tachometer is an angular speed sensing instrument. It is used in a variety of applications but only to monitor angular velocity.

Tachometer Conditioning

The tachometer used in the example is a dc tachometer. When the shaft of the tachometer moves, an output is generated. That output, although like dc, is analog. The magnitude of the voltage output is directly proportional to the field strength of the coil and the speed at which the armature rotates. The polarity of the signal is dependent on the polarity of the field and the direction of the rotating member. Specifications that are worth mentioning are normal, top, and low speeds, mounting configuration sizes, voltage at a specific rpm, linearity, accuracy, and temperature. Other major functions that are required for tachometer conditioning are listed below (see Figure 6–9). Refer to the common block descriptions at the beginning of this chapter for circuits flagged with asterisks (*).

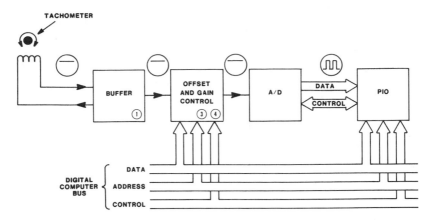

Figure 6–9 ANALOG-TO-DIGITAL CONVERSION: TACHOMETER (ANGULAR VELOCITY)

1. Isolation of the transducer from other circuitry
2. Amplification
3. Offset control
4. Gain control
5. Scaling control
*6. Analog-to-digital (A/D) conversion
*7. PIO interface with the bus

Buffer.   The buffer serves as a high-resistance input to the signal-conditioning circuit. The buffer prevents loading down of the transducer and thus distortion of the linearity of the transducer. The output of the transducer is fed into the buffer. The buffer isolates the transducer from other circuitry and presents it to the next stage as a voltage proportional to the measurand. Although the input to the buffer may be ac or dc from the transducer, it is still variable (analog).

Offset and gain control.   The offset and gain control circuit has several purposes. One of these purposes is to provide circuitry to eliminate error voltages imposed by the transducer. An offset voltage is applied in opposition to the error to ensure an input signal accuracy with zero error.

A second purpose of this stage is to adjust the amplification of the buffered voltage so that the voltage output is scaled properly. Scaling is required so that the signal is properly proportional to the variable being measured by the transducer.

The gain of the stage, as shown on the block diagram, is controlled automatically by the microprocessor/computer. This method is accomplished with the use of multiplying digital to analog converters and latches. A manual method may be provided which provides potentiometers for gain adjustment. The stage can also accommodate additional circuitry to implement a shunt resistor for calibration.

The offset function of this stage can be eliminated by programming the computer to measure the offset and store that value in memory. Each time a measurement is made, the computer subtracts the offset value from the measurand value to achieve the accurate variable.

A more complete implementation of programming method uses the straight-line formula

$$y = mx + b$$

This allows the computer to calculate the output by storing the offset value b and scale factor m. The measured variable y is then calculated for every buffered voltage x.

A final function of the offset gain control stage is to provide electronic circuitry for filtering noise. Software noise elimination may also be present. Software may be written by performing standard deviation calculations on the buffered voltage inputs.

The input to the offset and gain control is variable (analog). Output from the offset and gain control is directed to the analog-to-digital (A/D) converter. The output is direct current.

Excitation DC power supply.   Note that the tachometer is self-generating. Therefore, there is no excitation dc power supply. Although not shown on the block diagram, an appropriate power supply is necessary to establish dc power for electronics within the various blocks.

CAPACITANCE TRANSDUCERS

The capacitive transducer consists of two fixed conductive plates which are isolated from a housing by insulated standoffs. A pressure port directs pressure into a bellows. Attached to the bellows is a diaphragm. As pressure changes, the bellows expands or retracts, changing the position of the diaphragm, which serves as one capacitor plate. This causes a capacitance change in two separate capacitive circuits. In some transducer designs, only a single capacitive element is employed. The change in capacitance is used to vary the frequency of oscillators or to null a capacitance bridge. The dual-concept capacitive transducer allows two oscillators to operate in a highly linear mode. Small displacement of the capacitor diaphragm is a major advantage. When used in an oscillator circuit the output detected may be ac, dc, digital or phase shift. The capacitor transducer may be extremely small and operate at high temperatures.

The capacitive transducer is utilized in various measurement applications, including the following:

1. Linear displacement
2. Linear position

3. Linear acceleration
4. Angular displacement
5. Force
6. Sound
7. Pressure
8. Humidity

Capacitive Conditioning

The signal derived from a capacitive is produced by the dynamic variable changing the electrical capacitance of the transducer. The transducer then acts as a variable capacitor. The transducer forms part of the tuning capacitance of the oscillator of which it is an integral part. The oscillator in the example of Figure 6–10 is a Weinbridge type. The transducer is paired with another capacitor. As the transducer changes capacitance, the frequency of the oscillator changes.

The capacitive transducer may have accuracies within 0.1% even though they are inexpensive. The capacitive transducer has excellent frequency response, standard hysteresis, repeatability, and stability with excellent resolution. Its disadvantages include a high-impedance output with additional electronics. Capacitor leads from the sensor must be short to eliminate stray pickup. Finally, the capacitive transducer is extremely sensitive to temperature variations. Major functions necessary to signal condition the capacitive transducer are listed below (see Figure 6–10). Refer to the common block descriptions at the beginning of this chapter for circuits flagged with asterisks (*).

1. Isolation of the transducer from other circuitry by being part of a bridge within the oscillator
2. Oscillator
3. Frequency-to-voltage converter

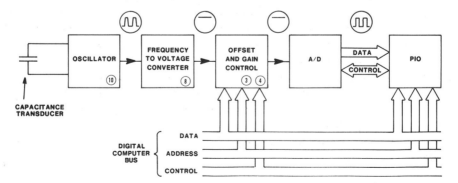

Figure 6–10 ANALOG-TO-DIGITAL CONVERSION: CAPACITANCE TRANSDUCER

4. Offset control
5. Gain control
6. Scaling control
*7. Analog-to-digital (A/D) conversion
*8. PIO interface with the bus

Oscillator.   The purpose of the oscillator in the capacitive transducer circuit is to provide circuitry to produce an alternating current that changes frequencies as the capacitance of the transducer changes. The transducer is part of the Weinbridge oscillator tuning circuit.

Frequency-to-voltage converter.   The purpose of the frequency-to-voltage converter is to convert the piezoelectric transducer frequency output into a direct-current (dc) voltage. This voltage is proportional to the transducer output. The input impedance of the frequency-to-voltage converter must be matched to the piezoelectric transducer so as to load the transducer properly and not distort its output signal. Impedance is provided by the transducer manufacturer. Conversion is usually done with a charge amplifier. This amplifier consists of a charge converter and a voltage amplifier. The charge converter is an operational amplifier with an integrating feedback. Some low-impedance output transducers come with electronics built into the case. These provide some amplification but low impedance. Their advantage is low noise; their disadvantage is nonadjustable gain.

The frequency-to-voltage converter used with the capacitive transducer is used to convert the frequency of the oscillator into a direct current proportional to that frequency. This requires a proper selection of resistors and capacitors so as to operate in the frequency range of the oscillator.

Offset and gain control.   The offset and gain control circuit has several purposes. One of these is to provide circuitry to eliminate error voltages imposed by the transducer. An offset voltage is applied in opposition to the error to ensure an input signal accuracy with zero error.

A second purpose of this stage is to adjust the amplification of the buffered voltage so that the voltage output is scaled properly. Scaling is required so that the signal is properly proportional to the variable being measured by the transducer.

The gain of the stage, as shown on the block diagram, is controlled automatically by the microprocessor/computer. This method is accomplished with the use of multiplying digital-to-analog converters and latches. A manual method may be provided which provides potentiometers for gain adjustment. The stage can also accommodate additional circuitry to implement a shunt resistor for calibration.

The offset function of this stage can be eliminated by programming the computer to measure the offset and store that value in memory. Each time a measurement is made, the computer subtracts the offset value from the measurand value to achieve the accurate variable.

A more complete implementation of programming method uses the straight-line formula

$$y = mx + b$$

This allows the computer to calculate the output by storing the offset value b and scale factor m. The measured variable y is then calculated for every buffered voltage x.

A final function of the offset gain control stage is to provide electronic circuitry for filtering noise. Software noise elimination may also be present. Software may be written by performing standard deviation calculations on the buffered voltage inputs.

The input to the offset and gain control is variable (analog). Output from the offset and gain control is directed to the analog-to-digital (A/D) converter. The output is direct current.

Excitation for oscillator and other circuits.   Note there is no power applied to the capacitive transducers since it is an integral part of the oscillator. Although not shown on the block diagram, an appropriate power supply is necessary to establish power for the oscillator and electronics within the various blocks.

MAGNETIC PICKUP TRANSDUCERS

The *turbine flow meter* consists of a bladed rotor suspended in a flow stream with its axis of rotation perpendicular to the flow direction. It is a velocity-measuring device calibrated to indicate volume and/or flow of liquid or gas in a pipe. The

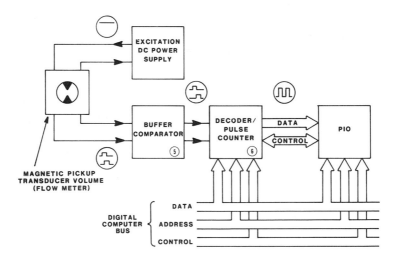

Figure 6–11   DIGITAL-TO-DIGITAL CONVERSION: MAGNETIC PICKUP TRANS-DUCER (VOLUME)

freely supported rotor revolves at a rate that is directly proportionally to the flow of the medium.

A *magnetic pickup* is used to sense the speed of a rotor. The pickup consists of a permanent magnet and a pickup coil. The pickup is mounted so that the rotor blades will cut the magnetic field. The rotor blades are made from a magnetic material so that as each blade cuts the field, a pulse is induced in the pickup coil. The output signal is a continuous sine-wave pulse train, with each pulse representing a discrete volume of the flowing medium. These pulses are fed to appropriate electronic units for volume or flow calculations.

When used for volume measurement (Figure 6–11) the number of pulses counted represent the distance the transducer has moved, which can be related to a volume by scaling techniques. When used for flow measurement, the frequency of the pulses represent the velocity of the transducer. Magnetic pickup transducers are used in flow meter applications.

Magnetic Pickup Transducer Conditioning

Signal generation uses a *magnetic pickup coil* which contains a permanent magnet and coil windings. The rotor or armature is located in proximity to the coil. Each time the rotor passes by the pickup coil, it induces one cycle of voltage in the coils. The frequency of the generated pulses is a quantity representing flow rate. The total number of pulses is a quantity representing total flow (volume).

Magnetic pickup flow meters are selective over a very large range. It is, however, difficult to find flow meters that will measure ranges from extremely low flows (less than 1 gpm) to high flows because of the geometrics involved. Specifications for magnetic pickup flow meters include accuracy; fluid densities; viscosity, flow velocity, temperature, and volume profiles; and mechanical lashup, including shear strengths.

Conditioning for volume applications.   Major functions necessary to signal condition the magnetic pickup transducer for volume applications (see Figure 6–11) are listed below. Refer to the common block description at the beginning of this chapter for the PIO description marked with an asterisk (*).

1. Excitation of the transducer and circuits
2. Isolation of the transducer from other circuitry
3. Amplification
4. Comparator
5. Decoder/pulse counter
*6. PIO interface with the bus

Buffer Comparator.   The buffer serves as a high-resistance input to the signal-conditioning circuit. The buffer prevents loading down of the transducer and thus distortion of its linearity. The output of the transducer, an analog voltage, is fed into the buffer comparator. The buffer isolates the encoder from other circuitry and presents it to the next stage as a voltage proportional to the measurand. The input to the buffer is variable (analog). The comparator is useful to determine whether a signal is above or below threshold. Its output saturates at the highest positive level, providing logic-level outputs. Further, the comparator "cleans up" the leading and falling edges of the transducer's pulse signals.

Decoder/Pulse Counter.   If the transducer has a pulse output, a decoder is used to convert the buffer output into a digital word that is representative of the transducer's position capacity. The other part of this stage is a pulse counter. That circuit counts buffer output pulses. The output of the pulse counter is a digital word representing the number of pulses counted. Since the output of the decoder/pulse counter is digital, there is no requirement for an A/D converter. Therefore, the decoder is connected directly to the PIO.

Conditioning for flow applications.   Major functions necessary to signal condition the magnetic pickup transducer for flow applications (see Figure 6–12) are listed below. Refer to the common block description at the beginning of this chapter for the PIO description marked with an asterisk (*).

1. Excitation of the transducer and circuits
2. Isolation of the transducer from other circuitry
3. Amplification
4. Comparator

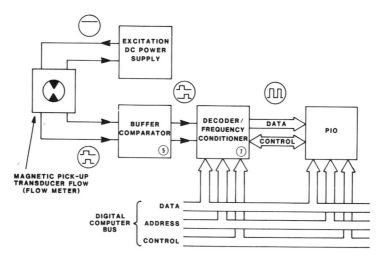

Figure 6–12   DIGITAL-TO-DIGITAL CONVERSION: MAGNETIC PICKUP TRANS-
DUCER (FLOW)

5. Decoder/frequency conditioner

*6. PIO interface with the bus

Buffer Comparator. The buffer serves as a high-resistance input to the signal-conditioning circuit. The buffer prevents loading down of the transducer and thus distortion of its linearity. The output of the transducer, an analog voltage, is fed into the buffer comparator. The buffer isolates the encoder from other circuitry and presents it to the next stage as a voltage proportional to the measurand. The input to the buffer is variable (analog). The comparator is useful to determine whether a signal is above or below threshold. Its output saturates at the highest positive level, providing logic-level outputs. Further, the comparator "cleans up" the leading and falling edges of the transducer's pulse signals.

Decoder/Frequency Conditioner. If the transducer has a pulse output, a decoder is used to convert the buffer output into a digital word that is representative of the transducer's frequency (flow). The other part of this stage is a frequency conditioner. This circuit is used to measure the frequency of output pulses of the buffer. The output of the frequency conditioner is a digital word representing the frequency of the buffer output pulses.

Excitation DC Power Supply. The power supply provides dc power to the magnetic pickup. Although not shown on the block diagram, an appropriate power supply is necessary to establish dc power for electronics within the various blocks.

THERMOCOUPLES

Thomas J. Seebeck discovered the thermocouple in 1823. This is also called the *Seebeck effect.* Seebeck fused two metal wires together on both their ends. He then heated one of the junctions and found that electron current flowed from one wire to the other. In this case, electron flow was from the copper wire to the iron wire. The heated junction is called the *hot junction*; the other is called the *cold junction.*

The potential developed across the heated junction is the thermocouple potential. Its polarity and magnitude are dependent on the type of material in the two dissimilar metals. Thermocouple transducer applications are obviously related to those involving temperature.

Thermocouple Conditioning

Thermocouples are chosen for their ability to provide a uniform voltage–temperature relationship. When temperature changes, the thermocouple should produce a linear change in voltage output. Some thermocouples operate well at high temperatures; others operate best at low temperatures. Some are not subject to corrosion, humidity, or oxidation. Others may be contaminated by exposure to specific elements. The type of thermocouple chosen must meet the requirements of the job. Since this is the case, standards have been developed by industry and the National Bureau of Standards. *Thermocouples* are designated by letter types. The common metals (called base materials) are ANSI (American National Standards Institute) types T, E, J, and K. The more exotic metals used are called noble metals. These are more expensive but are able to operate at higher temperatures and have high resistance to oxidation and corrosions.

The designations do require some explanation, because ANSI also provides letter designations for alloy types. Table 6–1 supplies the ANSI letter designation for alloys versus the alloy trade name. Table 6–2 lists the ANSI type and the metal–alloy combinations used to manufacture the thermocouple. After the material there are (+) and (−) signs. The (+) polarity establishes the metal with the higher-energy state. Signal inputs from the thermocouple are measured as using

TABLE 6–1   ANSI Designations Versus Trade Names of Alloys

ANSI Designation	Alloy (Generic or Trade Name)
JN, EN, or TN	Constantan, Cupron, Advance
JP	Iron
KN	Alumel, Nial T2, Thermokanthal KN
KP or EP	Chromel, tophel T1, Thermokanthal KP
RN or SN	Pure platinum
RP	Platinum 13% rhodium
SP	Platinum 10% rhodium
TP	Copper

TABLE 6–2   ANSI Symbol and Its Thermocouple Alloys

ANSI Symbol	Thermocouple Alloy
T	Copper$^{(+)}$ versus constantan$^{(-)}$
E	Chromel$^{(+)}$ versus constantan$^{(-)}$
J	Iron$^{(+)}$ versus constantan$^{(-)}$
K	Chromel$^{(+)}$ versus alumel$^{(-)}$
G*	Tungsten$^{(+)}$ versus tungsten 26% rhenium$^{(-)}$
C*	Tungsten 5% rhenium$^{(+)}$ versus tungsten 26% rhenium$^{(-)}$
R	Platinum$^{(+)}$ versus platinum 13% rhodium$^{(-)}$
S	Platinum$^{(+)}$ versus platinum 10% rhodium$^{(-)}$
B	Platinum 6% rhodium$^{(+)}$ versus platinum 30% rhodium$^{(-)}$

* These letters are not ANSI symbols.

a voltmeter at the differential amplifier and buffer. Major functions necessary to signal condition the thermocouple are listed below (see Figure 6–13). Refer to the common block descriptions at the beginning of this chapter for circuits flagged with asterisks (*).

1. Excitation of the electronic circuits
2. Isolation of the transducer from other circuitry
3. Amplification
4. Offset control
5. Gain control
6. Scaling control
*7. Analog-to-digital (A/D) conversion
*8. PIO interface with the bus

Differential amplifier and buffer.   The differential amplifier provides the proper electronics to measure the voltage across the output of the transducer. The buffer serves as a high-resistance input to the signal conditioning circuit. It also prevents loading down of the transducer and thus distortion of its linearity. The buffer isolates the transducer from other circuitry and presents it to the next stage as a voltage proportional to the measurand. The input to the buffer is analog (variable).

Offset and gain control.   The offset and gain control circuit has several purposes. One of these purposes is to provide circuitry to eliminate error voltages imposed by the transducer. An offset voltage is applied in opposition to the error to ensure an input signal accuracy with zero error.

A second purpose of this stage is to adjust the amplification of the buffered voltage so that the voltage output is scaled properly. Scaling is required so that the signal is properly proportional to the variable being measured by the transducer.

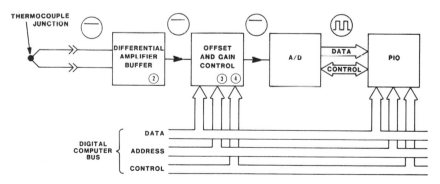

Figure 6–13 ANALOG-TO-DIGITAL CONVERSION: THERMOCOUPLE (TEMPERATURE)

The gain of the stage, as shown on the block diagram, is controlled automatically by the microprocessor/computer. This method is accomplished with the use of multiplying digital-to-analog converters and latches. A manual method may be provided which provides potentiometers for gain adjustment. The stage can also accommodate additional circuitry to implement a shunt resistor for calibration.

The offset function of this stage can be eliminated by programming the computer to measure the offset and store that value in memory. Each time a measurement is made, the computer subtracts the offset value from the measurand value to achieve the accurate variable.

A more complete implementation of programming method uses the straight-line formula

$$y = mx + b$$

This allows the computer to calculate the output by storing the offset value b and scale factor m. The measured variable y is then calculated for every buffered voltage x.

A final function of the offset gain control stage is to provide electronic circuitry for filtering noise. Software noise elimination may also be present. Software may be written by performing standard deviation calculations on the buffered voltage inputs.

The input to the offset and gain control is variable (analog). Output from the offset and gain control is directed to the analog-to-digital (A/D) converter. The output is direct current.

Excitation DC power supply.   Note that there is no power used to excite the thermocouple, as it is self-generating. Although not shown on the block diagram, an appropriate power supply is necessary to establish dc power for electronics within the various blocks.

THERMISTORS

Michael Faraday, an English scientist, found through experimentation that certain semiconductor materials decrease their resistance as temperature increases. The material is said to have a *negative temperature coefficient*. It was found later that oxides of cobalt, manganese, and nickel provide thermally sensitive resistance for temperature-involved applications. The resistance became known as the *thermistor*. The thermistor is a solid-state device that decreases in resistance as temperature increases. The word "thermistor" is derived from two words: *thermal* and *resistor*. In a circuit, the decrease in resistance also means an increase in current flow. Thermistors are extremely sensitive. Some may decrease in resistance as much as 5% for each degree (Celsius) rise in temperature. Thermistors are made from metallic oxide crystals by means of the *sintering process*.

A thermistor is installed in a circuit. Heat is applied to the thermistor. The thermistor decreases resistance. The potential across the thermistor decreases. When heat is taken away, the resistance increases and the potential across the thermistor increases. The thermistor has, then, sensed the change in heat so that it can be monitored as a signal. Obviously, thermistors applications are related to those involving temperature.

Thermistor Conditioning

Thermistors offer sensitivity to temperature differences. Although thermistors are nonlinear, they offer extreme sensitivity to small temperature changes.

Thermistors offer significant advantages in terms of matching impedance levels to available instrumentation or compensation circuit needs. For example, it is possible to choose a sensor with a 4% per °C or greater sensitivity in resistance levels of 100 Ω to 30 MΩ at 25°C. Thermistors offer significant fabrication advantages because they may be mounted in a great variety of substrates. Repeatability of measurements in a range of 0.001°C is easily achievable with long-term reproducibility of 0.005°C. Short-term reproducibility is frequently so good that measurement instrument error is the primary source of any uncertainty.

The principal disadvantages of thermistors for measurement applications are significant sensitivity changes and significant nonlinearity in absolute resistance sensitivity per degree. Specifications must include sensitivity, accuracy, temperature span, resistance values at temperature extremes, tolerances, and geometry.

The signal output from the thermistor is proportional to the temperature. Since the thermistor is connected directly to the differential amplifier, the amplifier feels any change. As the dynamic variable (temperature) is raised or lowered, the thermistor resistance varies accordingly. That change is felt at the amplifier. Major functions necessary to signal condition the thermistor are listed below (see Figure 6–14). Refer to the common block descriptions at the beginning of this chapter for circuits flagged with asterisks (*).

1. Excitation for the thermistor and circuits
2. Isolation of the thermistor from other circuitry
3. Amplification
4. Offset control
5. Gain control
6. Scaling control
*7. Analog-to-digital (A/D) conversion
*8. PIO interface with the bus

Differential amplifier and buffer.   The differential amplifier provides the proper electronics to measure the voltage across the output of the transducer. The buffer serves as a high-resistance input to the signal-conditioning circuit. It also prevents loading down of the transducer and thus distortion of its linearity.

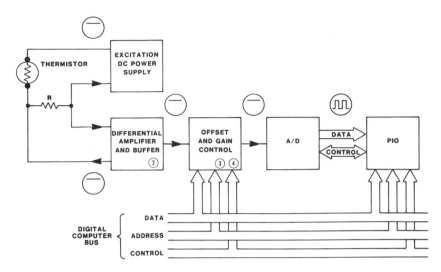

Figure 6–14 ANALOG-TO-DIGITAL CONVERSION: THERMISTOR (TEMPERA-TURE)

The buffer isolates the transducer from other circuitry and presents it to the next stage as a voltage proportional to the measurand. The input to the buffer is analog (variable).

Offset and gain control. The offset and gain control circuit has several purposes. One of these purposes is to provide circuitry to eliminate error voltages imposed by the transducer. An offset voltage is applied in opposition to the error to ensure an input signal accuracy with zero error.

A second purpose of this stage is to adjust the amplification of the buffered voltage so that the voltage output is scaled properly. Scaling is required so that the signal is properly proportional to the variable being measured by the transducer.

The gain of the stage, as shown on the block diagram, is controlled automatically by the microprocessor/computer. This method is accomplished with the use of multiplying digital-to-analog converters and latches. A manual method may be provided which provides potentiometers for gain adjustment. The stage can also accommodate additional circuitry to implement a shunt resistor for calibration.

The offset function of this stage can be eliminated by programming the computer to measure the offset and store that value in memory. Each time a measurement is made, the computer subtracts the offset value from the measurand value to achieve the accurate variable.

A more complete implementation of programming method uses the straight-line formula

$$y = mx + b$$

This allows the computer to calculate the output by storing the offset value b and scale factor m. The measured variable y is then calculated for every buffered voltage x.

A final function of the offset gain control stage is to provide electronic circuitry for filtering noise. Software noise elimination may also be present. Software may be written by performing standard deviation calculations on the buffered voltage inputs.

The input to the offset and gain control is variable (analog). Output from the offset and gain control is directed to the analog-to-digital (A/D) converter. The output is direct current.

Excitation DC power supply.   The dc power supply provides excitation for the thermistor and other circuits. A special resistor (R) is a current limiter. It is used to protect the thermistor from overcurrent and set up a voltage divider.

RESISTANCE TEMPERATURE DETECTORS

You may recall that the thermistor has a negative temperature coefficient. That is, as temperature increases, the resistance of the thermistor decreases. Most conductors of electricity such as the copper wire have a positive temperature ceofficient. These conductors increase in resistance as temperature increases. The positive temperature cocfficient is termed alpha (α). The thermal component that industry uses is the *resistance temperature detector*, known as the RTD.

The property of an RTD that is characteristic is its electrical resistance as a function of temperature. This term is known as alpha (α). (Refer to Table 6–3 for typical alpha temperature coefficients of RTD materials.) RTDs operate in the temperature range -400 to $+1700°F$. The RTD is more efficient than other temperature sensors in that their response to temperature is more linear. A change in temperature will provide an equivalent change in resistance over a long range of temperatures. The best of the RTDs is the platinum RTD. It has become a world standard in laboratory form for measurements between -270 and $+660°C$. Precautions and compromises encountered in using other types of electrical temperature sensors are unnecessary. Ordinary copper wire is used to connect the sensor to the readout instrument. Since the calibration is absolute, cold-junction compensation

TABLE 6–3   RTD Alpha Coefficients

Alpha Coefficient	RTD Material
0.0038	Copper
0.0039	Platinum
0.0045	Tungsten
0.0067	Nickel

is not necessary. The linear response eliminates corrective networks and errors in interpretation. Freedom from drift makes frequent recalibration unnecessary. Obviously, RTD applications are related to those involving temperature.

RTD Conditioning

The heart of a typical RTD is the sensing element. The sensing element is carefully stress-relieved and immobilized against strain or damage. The standard sensing element is mounted within a stainless-steel sheath in a manner that provides good thermal transfer and protection against moisture and the process medium. Sheaths are pressure-tight and may often be inserted directly into the process without thermowells. Sheaths are made of stainless steel, silver-soldered or heliarc-welded.

Each RTD is made of different materials and require different specifications. All RTDs must include sensitivity, accuracy, nominal resistance, material type and grade, temperature range, stability, response, tolerances, and geometry. The signal output from an RTD is proportional to the temperature. Since the RTD is connected directly to the differential amplifier, the amplifier feels any change. As the dynamic variable (temperature) is raised or lowered, the RTD varies accordingly. That change is felt at the amplifier.

Major functions necessary to signal condition the RTD are listed below (see Figure 6–15). Refer to the common block descriptions at the beginning of this chapter for circuits flagged with asterisks (*).

1. Excitation for the RTD and circuits
2. Isolation of the RTD from other circuitry
3. Amplification
4. Offset control
5. Gain control
6. Scaling control
*7. Analog-to-digital (A/D) conversion
*8. PIO interface with the bus

Differential amplifier and buffer. The differential amplifier provides the proper electronics to measure the voltage across the output of the transducer. The buffer serves as a high-resistance input to the signal-conditioning circuit. It also prevents loading down of the transducer and thus distortion of its linearity. The buffer isolates the transducer from other circuitry and presents it to the next stage as a voltage proportional to the measurand. The input to the buffer is analog (variable).

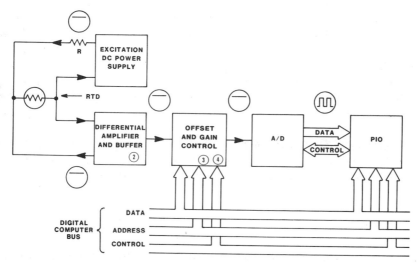

Figure 6-15 ANALOG-TO-DIGITAL CONVERSION: RESISTANCE TEMPERA-TURE DETECTOR (RTD) (TEMPERATURE)

Offset and gain control. The offset and gain control circuit has several purposes. One of these purposes is to provide circuitry to eliminate error voltages imposed by the transducer. An offset voltage is applied in opposition to the error to ensure an input signal accuracy with zero error.

A second purpose of this stage is to adjust the amplification of the buffered voltage so that the voltage output is scaled properly. Scaling is required so that the signal is properly proportional to the variable being measured by the transducer.

The gain of the stage, as shown on the block diagram, is controlled automatically by the microprocessor/computer. This method is accomplished with the use of multiplying digital-to-analog converters and latches. A manual method may be provided which provides potentiometers for gain adjustment. The stage can also accommodate additional circuitry to implement a shunt resistor for calibration.

The offset function of this stage can be eliminated by programming the computer to measure the offset and store that value in memory. Each time a measurement is made, the computer subtracts the offset value from the measurand value to achieve the accurate variable.

A more complete implementation of programming method uses the straight-line formula

$$y = mx + b$$

This allows the computer to calculate the output by storing the offset value b and scale factor m. The measured variable y is then calculated for every buffered voltage x.

A final function of the offset gain control stage is to provide electronic circuitry for filtering noise. Software noise elimination may also be present. Software may be written by performing standard deviation calculations on the buffered voltage inputs.

The input to the offset and gain control is variable (analog). Output from the offset and gain control is directed to the analog-to-digital (A/D) converter. The output is direct current.

Excitation DC power supply.   The dc power supply provides excitation for the RTD and other circuits. A special resistor (R) is a current limiter. It is used to protect the RTD from overcurrent and set up a voltage divider.

PHOTODETECTION

The most basic form of photodetection is the *photoresistor*. The photoresistor is a small slice of photoconductive material whose resistance decreases or increases as light energy is applied. Electrons are released by the light and flow toward a positive power supply. The basic task of the photoresistor is to convert light energy to electrical energy. The photoresistive material is nonreflective. There are no junctions in a bulk photoresistor.

A second detector is the single-junction *photodiode*. A photodiode is the optical version of the standard diode. It is constructed of a *pn* junction. Photons of light energy are absorbed into the device. Hole–electron pairs are generated. The pairs are combined at different depths within the diode depending on the energy level of the photon. A wide, thin surface area is used to ensure maximum absorption. Current flow is dependent on the amount of radiation that is absorbed.

Photodiodes operate with and without dc bias. The solar cell is a photodiode that is heavily doped. The depletion area is extremely thin, as is its radiation area. The cell is coated to avoid reflection. Hole–electron pairs diffuse to the depletion area of the diode, where they are drawn out as useful current. Output current is dependent on input radiation. Solar cells are not biased and are photovoltaic in operation.

Phototransistors are two-junction devices that have a large base area. The base region of the phototransistor absorbs the photons of energy and generates hole–electron pairs in the large base–collector region. The collector, being reverse-biased, draws the holes toward the base and the electrons toward the collector. The forward-biased base–emitter junction causes holes to flow from base to emitter and electrons to flow from emitter to base. Forward bias causes the phototransistor to operate just as the conventional transistor operates. The basic function, then, is that the light energy induces the transistor to conduct. Obviously, photodetection device applications are related to those involving light.

Photodetector Conditioning

The basic premise of a photodetector is that whichever type is used, they all respond to light. The resistor changes its resistance. The photodiode begins or changes its conduction. The phototransistor begins or changes its conduction. All three types are placed in circuitry that connects to a differential amplifier.

Signal output from the photoresistor, photodiode, and phototransistor are proportional to the dynamic variable of light. Since the photodetectors are connected directly to the differential amplifier, the amplifier feels any change. As light increases or decreases, the photodetector responds. That response is felt at the differential amplifier.

Specifications include spectral response curves, intensity, operating voltage and current, temperature, and in general all those related functions of resistors, diodes, and transistors. Major functions necessary to signal condition the photodetectors are listed below (see Figures 6–16, 6–17, and 6–18). Refer to the common block descriptions at the beginning of this chapter for circuits flagged with asterisks (*).

1. Excitation for the photodetectors and circuits
2. Isolation of the photodetectors from other circuitry
3. Amplification
4. Offset control

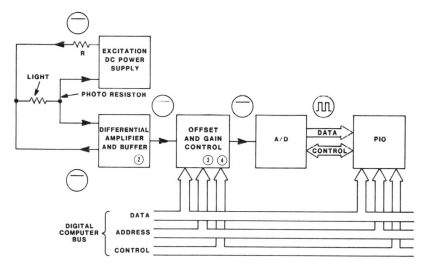

Figure 6–16 ANALOG-TO-DIGITAL CONVERSION: PHOTORESISTOR (LIGHT)

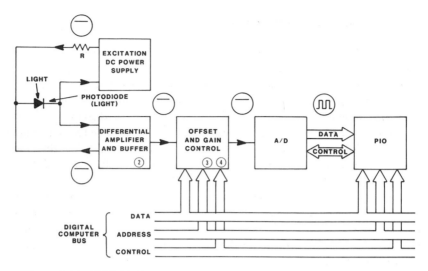

Figure 6–17   ANALOG-TO-DIGITAL CONVERSION: PHOTODIODE (LIGHT)

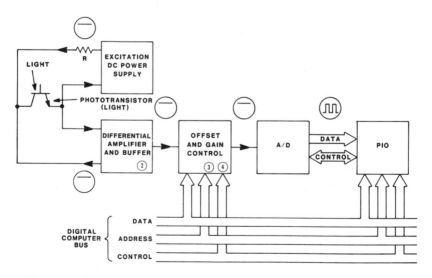

Figure   6–18   ANALOG-TO-DIGITAL   CONVERSION: PHOTOTRANSISTOR
(LIGHT)

5. Gain control

6. Scaling control

*7. Analog-to-digital (A/D) conversion

*8. PIO interface with the bus

Differential amplifier and buffer. The differential amplifier provides the proper electronics to measure the voltage across the output of the transducer. The buffer serves as a high-resistance input to the signal-conditioning circuit. It also prevents loading down of the transducer and thus distortion of its linearity. The buffer isolates the transducer from other circuitry and presents it to the next stage as a voltage proportional to the measurand. The input to the buffer is analog (variable).

Offset and gain control. The offset and gain control circuit has several purposes. One of these purposes is to provide circuitry to eliminate error voltages imposed by the transducer. An offset voltage is applied in opposition to the error to ensure an input signal accuracy with zero error.

A second purpose of this stage is to adjust the amplification of the buffered voltage so that the voltage output is scaled properly. Scaling is required so that the signal is properly proportional to the variable being measured by the transducer.

The gain of the stage, as shown on the block diagram, is controlled automatically by the microprocessor/computer. This method is accomplished with the use of multiplying digital to analog converters and latches. A manual method may be provided which provides potentiometers for gain adjustment. The stage can also accommodate additional circuitry to implement a shunt resistor for calibration.

The offset function of this stage can be eliminated by programming the computer to measure the offset and store that value in memory. Each time a measurement is made, the computer subtracts the offset value from the measurand value to achieve the accurate variable.

A more complete implementation of programming method uses the straight-line formula

$$y = mx + b$$

This allows the computer to calculate the output by storing the offset value b and scale factor m. The measured variable y is then calculated for every buffered voltage x.

A final function of the offset gain control stage is to provide electronic circuitry for filtering noise. Software noise elimination may also be present. Software may be written by performing standard deviation calculations on the buffered voltage inputs.

The input to the offset and gain control is variable (analog). Output from the offset and gain control is directed to the analog-to-digital (A/D) converter. The output is direct current.

Excitation DC power supply.   The dc power supply provides excitation for the photodetector and other circuits. A special resistor (R) is a current limiter. It is used to protect the photodetectors from overcurrent and set up a voltage divider.

HALL EFFECT POWER TRANSDUCERS

The *Hall effect* was discovered by E. H. Hall in 1879. The Hall effect, for purposes of this discussion, is that characteristic of a certain crystal such that when it is conducting current (control current) and is placed in a magnetic field, a potential difference is produced across its opposite edges. The potential difference is proportional to the product of the control current, the strength of the field, and the cosine of the phase angle between the control current and magnetic flux. By putting the crystal in a magnetic structure so that an ac current generates the flux and an ac voltage produces the control current, we have a multiplying device. It produces an output proportional to IE cos (power). This is in essence the Hall generator watt transducer. The load voltage (E) produces a proportional control current (I) through the crystal. The output is proportional to the product of EI and the phase angle between them.

Actually, the output of the Hall generator consists of a dc voltage proportional to true power (watts) plus a double-frequency ac voltage proportional to the volt-amperes in the circuit. When the transducer is used with devices that do respond to ac, the double-frequency component must be filtered out. Filters are provided especially for this purpose.

The output of a Hall element decreases with increasing temperature. The watt transducers are temperature-compensated with a thermistor-resistor network in the output circuit. The load resistance is part of the network; therefore, it is fixed for a particular transducer. The Hall effect transducer can be used in alternating current, voltage, and/or power measurements applications.

Hall Effect Power Transducer Conditioning

The Hall effect power transducer outputs an alternating current that is proportional to the power consumption of the transducer. As the alternating current being monitored by the transducer changes, the output of the transducer changes. The transducer requires circuitry to filter out the alternating current component of the transducers output. Signal conditioning of the director current (dc) component is necessary because the dc component is proportional to the power being measured.

The Hall current or voltage transducer converts an ac current input into a proportional dc current of low magnitude. The transducer consists of a current

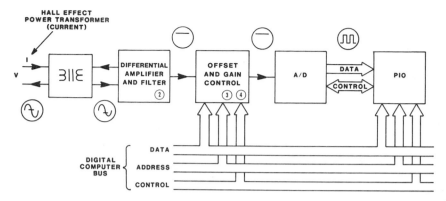

Figure 6–19 ANALOG-TO-DIGITAL CONVERSION: HALL EFFECT TRANS-
DUCER (CURRENT)

transformer that feeds through a calibrating rheostat to a full-wave rectifier. An
RC circuit in the network in the circuit is used for waveform error compensation.
On power systems a pure waveform is rare. Close to the source generator the
voltage tends to be clean but the current wave is distorted. Distortions are due to
harmonics created by equipment and parts attached to the system for distribution
and transformation.

Major functions necessary for conditioning the Hall effect transducer are
listed below (see Figure 6–19). Refer to the common block descriptions at the
beginning of this chapter for circuits flagged with asterisks (*).

1. Isolation of the transducer from other circuitry
2. Amplification
3. Filtering
4. Offset control
5. Gain control
6. Scaling control
*7. Analog-to-digital (A/D) conversion
*8. PIO interface with the bus

Differential amplifier and filter.   The differential amplifier and filter
are used with the Hall effect transducer. The differential amplifier provides the
proper electronics to measure the voltage across the output of the transducer.
The buffer serves as a high-resistance input to the signal-conditioning circuit. It
also prevents loading down of the transducer and thus distortion of its linearity.
The filter is a notch filter. Its purpose is to eliminate the alternating-current part
of the measured signal and allow the direct-current part to remain. The direct
current is then presented to the next stage to be conditioned.

Offset and gain control.   The offset and gain control circuit has several purposes. One of these purposes is to provide circuitry to eliminate error voltages imposed by the transducer. An offset voltage is applied in opposition to the error to ensure an input signal accuracy with zero error.

A second purpose of this stage is to adjust the amplification of the buffered voltage so that the voltage output is scaled properly. Scaling is required so that the signal is properly proportional to the variable being measured by the transducer.

The gain of the stage, as shown on the block diagram, is controlled automatically by the microprocessor/computer. This method is accomplished with the use of multiplying digital-to-analog converters and latches. A manual method may be provided which provides potentiometers for gain adjustment. The stage can also accommodate additional circuitry to implement a shunt resistor for calibration.

The offset function of this stage can be eliminated by programming the computer to measure the offset and store that value in memory. Each time a measurement is made, the computer subtracts the offset value from the measurand value to achieve the accurate variable.

A more complete implementation of programming method uses the straight-line formula

$$y = mx + b$$

This allows the computer to calculate the output by storing the offset value b and scale factor m. The measured variable y is then calculated for every buffered voltage x.

A final function of the offset gain control stage is to provide electronic circuitry for filtering noise. Software noise elimination may also be present. Software may be written by performing standard deviation calculations on the buffered voltage inputs.

The input to the offset and gain control is variable (analog). Output from the offset and gain control is directed to the analog-to-digital (A/D) converter. The output is direct current.

Excitation DC power supply.   Note that there is no excitation for the transducer because it is self-generating. Although not shown on the block diagram, an appropriate power supply is necessary to establish dc power for the PIO.

ON/OFF SWITCHES

There are probably more switch devices than any other electric device. They serve the purpose of turning things on and off. By doing this, control of a system is accomplished. Switches are functionally used to open or close a circuit. When a switch is ON, it is closed and in the *make* position. When a switch is OFF, it is open and in the *break* position. There are four basic switch functions that a switch makes: (1) maintained contact, (2) momentary contacts, (3) make before

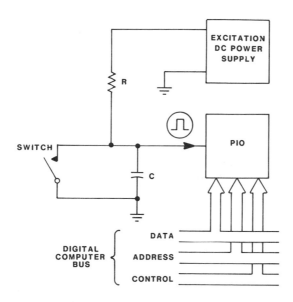

Figure 6–20   DIGITAL-TO-DIGITAL CONVERSION: SWITCHING

break, and (4) break before make. Switches are used to connect and disconnect power, to limit or curtail circuit action, to enable or disable circuitry, and to do many other things. Switches are built into components such as transducers. Some switches are lighted. Some are numbered. The shape of the switch depends on its application and the space available. Applications for the switch include everything that requires an on/off configuration.

On/Off Switch Conditioning

The major purpose of the switch in a computer-controlled system is to provide a means of telling the computer (microprocessor) when an event has taken place. The switch is in either the on or off configuration. Switches are mechanical devices and inherently "bounce" when they close. The resistor (R) and the capacitor (C) in Figure 6–20 provide a method of filtering the voltage oscillation when the switch bounces after closure. This eliminates nuisance pulses that could be picked up by a high-speed parallel input/output PIO stage. The switch in Figure 6–21 is a light-operated phototransistor. Like the mechanical switch, this switch is either on or off and requires only PIO connection to the bus. The major functions necessary to condition the switch are the input RC filter and the PIO interface to the bus. Refer to the common block description for the PIO at the beginning of this chapter.

Excitation DC power supply.   It is obvious that the switch does not require excitation. Although not shown on the block diagram, an appropriate power supply is necessary to establish dc power to the PIO interface block electronics.

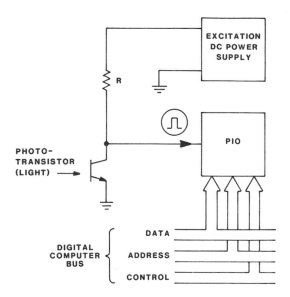

Figure 6–21 DIGITAL-TO-DIGITAL CONVERSION: PHOTOTRANSISTOR SWITCH (LIGHT)

PROXIMITY SWITCHES

The capacitance *proximity switch* is used for proximity detection. As an object comes close to the switch the electrical capacitance between the object and the switch changes and causes the switch to close. The proximity switch can be used in any application where it is electrically and mechanically compatible with the environment.

Proximity Switch Conditioning

Input to the switch is a change of environment around the switch. Proximity of an object to the switch will make or break the switch contacts. When the switch contacts change position, that information is fed to the buffer and comparator. This stage ensures that the capacitance switch is not loaded down. The comparator may require a certain amount of hysteresis in order to eliminate any possible oscillations during the switching process. The switching inputs to the PIO should be crisp to prevent nuisance tripping or interruptions within the computer. Functions necessary for proximity switch conditioning are listed below (see Figure 6–22). Refer to the common block description at the beginning of this chapter for the PIO description marked with an asterisk (*).

1. Excitation for the capacitance proximity switch
2. Isolation of the capacitance proximity switch from other circuitry

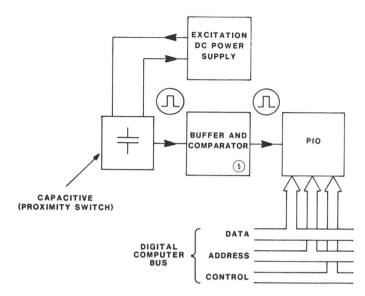

Figure 6–22 ANALOG-TO-DIGITAL CONVERSION: CAPACITANCE (PROXIM-ITY)

3. Amplifier
4. Comparator
*5. PIO interface with the bus

Buffer comparator.   The buffer serves as a high-resistance input to the signal-conditioning circuit. The buffer prevents loading down of the transducer and thereby distorting its linearity. The output of the transducer, an analog voltage, is fed into the buffer comparator. The buffer isolates the encoder from other circuitry and presents it to the next stage as a voltage proportional to the measurand. The input to the buffer is variable (analog). The comparator is useful to determine whether a signal is above or below threshold. Its output saturates at the highest positive level, providing logic-level outputs. Further, the comparator "cleans up" the leading and falling edges of the transducer's pulse signals.

Excitation DC power supply.   The capacitance proximity switch requires dc power for operation. Although not shown on the illustration, an appropriate power supply is required to energize electronic circuits in the various blocks.

7

Conditioning for Control Drivers

Transducers have analog signal outputs that must be converted to digital to be compatible to the bus. In the case of control drivers, their intelligence and operations are dictated by the microprocessor. Of course, the microprocessor's output is digital. Therefore, conversion from digital to analog is required. Signal-conditioning diagrams for control drivers in this chapter include many blocks which are common to several digital-to-analog (D/A) conversions. The purposes of these common blocks (circuits) are repeated as necessary to obviate searching by the reader. Within the block diagrams are small circled numbers. These numbers are related to an electronic circuit found in Appendix A. Power supplies are not included. They may be of any variety as long as they are compatible in power capability.

DIRECT-CURRENT SOLENOIDS

The solenoid has a ferromagnetic core that is wrapped circularly by a coil of wire. The coil is generally wound in layers. The core has a cylindrical opening. In the center of the core is a free-moving iron or steel rod or plunger. A steel plate is mounted just above the core. The core becomes highly magnetized when current is applied through the coils of wire. The polarized magnetic field caused by excitation of the coil asserts pressure on the free-moving rod in the center of the core, accelerating it through the hole in the core. The rod strikes the iron plate and closes contacts in the same manner that a switch closes. The solenoid can therefore be used to mechanically open or close mechanical fixtures such as a hydraulic valve. The solenoid is either on or off but can never be between or proportional. The solenoid is a basic method for converting electrical current to mechanical energy.

Conditioning to Drive Direct-Current Solenoids

The typical direct-current solenoid is an electromechanical device whose geometry and strength are as important as signal requirement. The designer must understand its force specifications of pull or push, stroke distance of the plunger, and timing to actuate or to deenergize. Ambient temperature range is another specification required. Of course, voltage and current ratings are necessary information. Functions required to condition the direct-current solenoid for mechanical operation are as follows (see Figure 7–1):

1. PIO interface from the bus
2. Decoding
3. Buffering
4. Latching
5. Power driving

Parallel Input/output.   The purpose of the PIO is to interface the computer bus directly with the setup and control circuitry and/or the decoder. The stage must provide the "handshaking" between the setup and control circuitry and the computer by controlling data bus traffic direction and signaling the control service to tell it when to turn on or off.

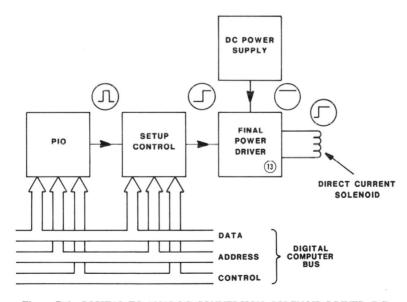

Figure 7–1 DIGITAL-TO-ANALOG CONVERSION: SOLENOID DRIVER (DC)

Generally, the PIO is a single IC and is part of the computer's IC chip set. The PIO recognizes when it is to act/react by address lines from that bus. The PIO performs actions in response to instructions from the control bus. Information is taken or given by way of the data bus. Input to the PIO is a digital word from the computer. Output from the PIO is a digital word to the setup and control circuitry and/or decoder.

A more detailed explanation of the PIO is found in Chapter 8 in the section "The Microprocessor." PIO integrated circuits usually accompany the microprocessor chip and therefore are described along with the microprocessor.

Setup control for direct-current applications.   The purpose of the setup control is to provide circuitry to decode and to determine which device is to be set into operation. The setup control also provides circuits to buffer the computer bus and latch in the information as to which way to control the device either on or off.

Final power drive.   The final power drive interfaces the setup control stage with the device to be controlled. This stage accepts a signal logic level of low power from the setup control stage and translates it into the power level required to operate the controlled device. The final power drive usually contains a power transistor switch, a triac, or a solid-state relay. In conjunction with functioned circuits, the final power drive must contain short-circuit protection for high-voltage spikes due to inductive surge. Circuits can be current limiters for power supply protection, along with diodes placed in parallel with the controlled device to eliminate spikes.

Excitation DC power.   The dc power supply provides adequate power to drive the controlled device (solenoid). It must contain current-limiting circuitry to protect itself and final power drive from overload or shorting. Although not shown on the block diagram, dc power is also required to energize the electronic circuits within the various blocks.

ALTERNATING CURRENT SOLENOIDS

Operation of the alternating-current solenoid is basically the same as the direct-current solenoid (refer to the description of the direct-current solenoid). Although alternating current operates the solenoid, the coils are mechanized and polarized so that the plunger operates in only one direction, regardless of which way power is applied to the ac power leads. As with the dc solenoid, this device is either on or off, never in between or proportional. The ac solenoid is a basic method for converting electrical current to mechanical energy.

Conditioning to Drive Alternating-Current Solenoids

The typical alternating-current solenoid is an electromechanical device whose geometry and strength are as important as signal requirement. The designer requires force specifications of push or pull, stroke distance of the plunger, and timing to actuate or to deenergize. Ambient temperature ranges are a specification worthy of consideration. Of course, voltage, current, and frequency ratings are necessary information, along with mechanical hookup requirements. Functions required to condition the alternating current solenoid for some mechanical operation are as follows (see Figure 7–2):

1. PIO interface to the bus
2. Decoding
3. Buffering
4. Latching
5. Zero-crossing detection
6. Power driving

Parallel input/output. The purpose of the PIO is to interface the computer bus directly with the setup and control circuitry and/or the decoder. The stage must provide the "handshaking" between the setup and control circuitry and the

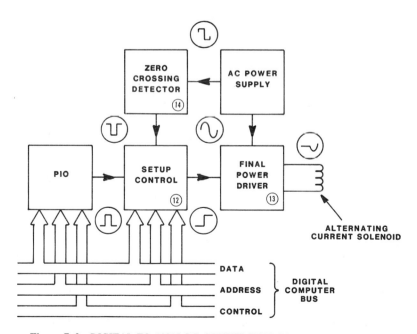

Figure 7–2 DIGITAL-TO-ANALOG CONVERSION: SOLENOID DRIVER (AC)

computer by controlling data bus traffic direction and signaling the control service to tell it when to turn on or off.

Generally, the PIO is a single IC and is part of the computer's IC chip set. The PIO recognizes when it is to act/react by address lines from that bus. The PIO performs actions in response to instructions from the control bus. Information is taken or given by way of the data bus. Input to the PIO is a digital word from the computer. Output from the PIO is a digital word to the setup and control circuitry and/or decoder.

A more detailed explanation of the PIO is found in Chapter 8 in the section "The Microprocessor." PIO integrated circuits usually accompany the microprocessor chip and therefore are described along with the microprocessor.

Setup control for alternating-current application.   The purpose of the setup control is to provide circuitry to decode and to determine which device is to be set in operation. The setup control also provides circuits to buffer the computer bus and latch in the information as to which way to control the device either on or off. A final circuit accepts the output of the zero-crossing detector. This function enables the control device to turn on only when the zero-crossing detector acknowledges the go-ahead. This action is required to eliminate a high current surge when the control device is turned on.

Final power drive.   The final power drive interfaces the setup control stage with the device to be controlled. This stage accepts a signal logic level of low power from the setup control stage and translates it into the power level required to operate the controlled device. The final power drive usually contains a power transistor switch, a triac, or a solid-state relay. In conjunction with functioned circuits, the final power drive must contain short-circuit protection for high-voltage spikes due to inductive surge. Circuits can be current limiters for power supply protection, along with diodes placed in parallel with the controlled device to eliminate spikes.

Zero-crossing detector.   The zero-crossing detector has one purpose. It provides an enable pulse each time the ac power supply current transitions through zero. This pulse is fed to the setup control for synchronization.

Alternating-current power supply.   The ac power supply energizes and provides current for the controlled device. It must contain enough capability to drive the number of controlled devices for which it is responsible. Further, it should provide current limiting to protect itself and the final power drive from burning up in the event of a short. Although not shown on the block diagram, dc power is required to energize the electronic circuits within the various blocks.

RELAYS

Relays are usually defined as electrically controlled switches. This is probably true in the real sense. Most relays are controlled by energizing an electrical coil, which, in turn, electromagnetically operates a movable slug and breaks or makes a set of contacts.

Most relays operate by the electromagnetic function. Power is applied to a coil. The coil energizes, and an armature within the coil is attracted by the electromagnetic force. The armature moves downward or upward and makes contact. The plunger-type relay has the armature within the relay coil. Power is applied to the coil energizing it. The coil is polarized with the armature. The armature is driven from the coil by electromagnetism and breaks contact.

The electromagnetic relay can have a great number of contacts. It is, however, slow to function, and the contacts may freeze up or burn off because of high current and arcing. Contacts also may chatter (open and close) before total closure takes place. The relay is a basic method for controlling single or multiple loops of current which may drive a single or multiple mechanisms.

Relay Conditioning

Input to the electromagnetic relay is dc current. Specifications required are geometrical considerations, weight, contact configurations, contact types and material, times for contact closing and opening, and temperature requirements. Of course, power, voltage, and current ratings are necessary information. Functions required to condition the dc relay for mechanical/electrical connection are as follows (see Figure 7–3):

1. PIO interface with the bus
2. Decoding
3. Buffering
4. Latching
5. Power driving

Parallel input/output.   The purpose of the PIO is to interface the computer bus directly with the setup and control circuitry and/or the decoder. The stage must provide the "handshaking" between the setup and control circuitry and the computer by controlling data bus traffic direction and signaling the control service to tell it when to turn on or off.

Generally, the PIO is a single IC and is part of the computer's IC chip set. The PIO recognizes when it is to act/react by address lines from that bus. The PIO performs actions in response to instructions from the control bus. Information is taken or given by way of the data bus. Input to the PIO is a digital word from

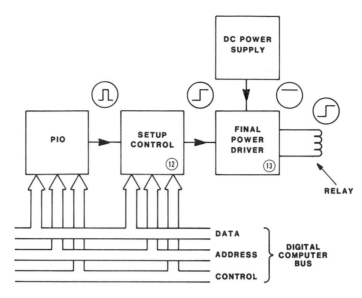

Figure 7–3 DIGITAL-TO-ANALOG CONVERSION: RELAY (DC)

the computer. Output from the PIO is a digital word to the setup and control circuitry and/or decoder.

A more detailed explanation of the PIO is found in Chapter 8 in the section "The Microprocessor." PIO integrated circuits usually accompany the microprocessor chip and therefore are described along with the microprocessor.

Setup control for direct-current applications. The purpose of the setup control is to provide circuitry to decode and to determine which device is to be set into operation. The setup control also provides circuits to buffer the computer bus and latch in the information as to which way to control the device either on or off.

Final power drive. The final power drive interfaces the setup control stage with the device to be controlled. This stage accepts a signal logic level of low power from the setup control stage and translates it into the power level required to operate the controlled device. The final power drive usually contains a power transistor switch, a triac, or a solid-state relay. In conjunction with functioned circuits, the final power drive must contain short-circuit protection for high-voltage spikes due to inductive surges. Circuits can be current limiters for power supply protection, along with diodes placed in parallel with the controlled device to eliminate spikes.

Excitation dc power. The dc power supply provides adequate power to drive the control device (relay). It must contain current-limiting circuitry to protect itself and final power drive from overload or shorting. Although not shown on the block diagram, dc power is also required to energize the electronic circuits within the various blocks.

SERVOMOTORS

The servomotor is essentially an ordinary dc motor. It does, however, have some special features. The servomotor is required to produce rapid accelerations from a standstill. Physical requirements for such a motor are lower inertia and high starting torque. Low inertia is obtained with the use of a small-diameter armature. The armature length is extended to improve power output. The servomotor may have a fixed field (armature controlled) or a fixed armature (field controlled) current.

The servomotor may be proportionally controlled or analog with infinite resolution. For each input there is an output. For example, if the servomotor were used as a flow control for a hydraulic system, a given amount of current input will provide a specific flow output.

The servomotor in Figure 7–4 is a single-ended motor. It has a single coil. Another servomotor used extensively with control systems is the push-pull (differential). This motor has a center-tapped coil. It has two inputs with the center grounded. Applications are extensive for single or dual inputs.

Servomotor Conditioning

Input to the servomotor is dc current. For a specified current a specific result output will occur. Specifications are similar to those of any dc motor. Some

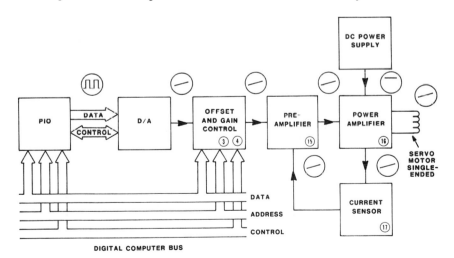

Figure 7–4 DIGITAL-TO-ANALOG CONVERSION: SERVOMOTOR (SINGLE-ENDED)

of these are torque, inertia, geometry, and temperature ranges. Power requirements along with voltage and current ratings are also significant. Functions required to condition the servomotor are as follows (see Figure 7–4):

1. PIO interface from the bus
2. Digital-to-analog (D/A) conversion
3. Offset control
4. Gain control
5. Filtering
6. Preamplification
7. Current sensing
8. Power amplifying

Parallel input/output.   The purpose of the PIO is to interface the computer bus directly with the D/A converter and drive circuitry. The stage must provide the "handshaking" between drive circuitry and the computer by controlling data bus traffic direction and signaling the control service to tell it when to turn on or off.

Generally, the PIO is a single IC and is part of the computer's IC chip set. The PIO recognizes when it is to act/react by address lines from that bus. The PIO performs actions in response to instructions from the control bus. Information is taken or given by way of the data bus. Input to the PIO is a digital word from the computer. Output from the PIO is a digital word to the setup and control circuitry and/or decoder.

A more detailed explanation of the PIO is found in Chapter 8 in the section "The microprocessor." PIO integrated circuits usually accompany the microprocessor chip and therefore are described along with the microprocessor.

Digital-to-analog converter.   The digital-to-analog (D/A) converter is a circuit that converts a digital signal (word) into an analog voltage or current that is proportional to the digital word. For example, an 8-bit D/A converter may output a signal of 0 V when an adjusted word of 00000000_2 is input. At the top of the scale, the same 8-bit D/A converter may output 10 V when a digital word of 11111111_2 is input.

In the case of the control device conditioning in this chapter, the digital word is input from the parallel input/output (PIO) into the D/A converter. This D/A converter was selected to interface with the microprocessor in data range, address range, and digital word characteristics.

Offset and gain control.   The purpose of the offset and gain control is to allow for microprocessor control of forward path offset and gain, which are part of the dynamic stability of the control device. A second purpose is filtering of spikes and other ac-like signals that might be riding the digital input. This stage is necessary for closed-loop servomotor or servomechanism systems.

Preamplifier.   The preamplifier provides additional amplification of the input voltage and makes the translation from voltage to current drive. The servomotor is a current-controlled device. To control it properly, the voltage-to-current conversion is necessary.

Current sensor.   The current sensor has circuitry which converts the current passing through the servomotor into a voltage. That voltage is fed back into the preamplifier to provide stability. The stage has a precision resistor in series with the servomotor. The voltage drop across the resistor is equivalent to the current flow through the resistor.

Power amplifier (single ended).   This stage translates the low power current coming from the preamplifier into a high power current to drive the servomotor. This amplifier is usually a power transistor or a complementary power transistor pair. The single-ended power amplifier drives a single coil in the servomotor or a pair of coils that are in a parallel-aided configuration.

Excitation dc power.   The dc power supply provides adequate power to drive the controlled device (servomotor). It must contain current-limiting circuitry to protect itself and final power drive from overload or shorting. Although not shown on the block diagram, dc power is also required to energize the electronic circuits within the various blocks.

DIRECT-CURRENT MOTORS

A dc motor converts electricity into rotary motion. A motor consists of a magnetic field with a rotating armature excited by an external source. Speed of the armature is dependent on the current applied. Torque is produced by back EMF. The field coil is placed in series with the armature or in shunt (parallel). In series the motor is hard to control but has a large starting torque. In parallel the motor is easy to control but has a low starting torque. Compound coil windings provide both series and parallel hookups. The dc motor provides a means of converting an electrical current into mechanical energy.

Direct-Current Motor Conditioning

The direct-current motor is a electromechanical device whose geometry and strength are as important as signal requirement. The designer must understand torque specifications and armature speeds along with ambient and working temperatures. Of course, voltage and current ratings are necessary information. Functions

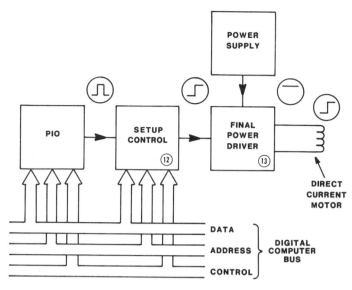

Figure 7-5   DIGITAL-TO-ANALOG CONVERSION: MOTOR (DC)

required to condition the direct-current motor for mechanical operation are as follows (see Figure 7–5):

1. PIO interface from the bus
2. Decoding
3. Buffering
4. Latching
5. Power driving

Parallel input/output.   The purpose of the PIO is to interface the computer bus directly with the setup and control circuitry and/or the decoder. The stage must provide the "handshaking" between the setup and control circuitry and the computer by controlling data bus traffic direction and signaling the control service to tell it when to turn on or off.

Generally, the PIO is a single IC and is part of the computer's IC chip set. The PIO recognizes when it is to act/react by address lines from that bus. The PIO performs actions in response to instructions from the control bus. Information is taken or given by way of the data bus. Input to the PIO is a digital word from the computer. Output from the PIO is a digital word to the setup and control circuitry and/or decoder.

A more detailed explanation of the PIO is found in Chapter 8 in the section "The Microprocessor." PIO integrated circuits usually accompany the microprocessor chip and therefore are described along with the microprocessor.

Setup control for direct-current applications.   The purpose of the setup control is to provide circuitry to decode and to determine which device is to be set into operation. The setup control also provides circuits to buffer the computer bus and latch in the information as to which way to control the device either on or off.

Final power drive.   The final power drive interfaces the setup control stage with the device to be controlled. This stage accepts a signal logic level of low power from the setup control stage and translates it into the power level required to operate the controlled device. The final power drive usually contains a power transistor switch, a triac, or a solid-state relay. In conjunction with functioned circuits, the final power drive must contain short-circuit protection for high-voltage spikes due to inductive surge. Circuits can be current limiters for power supply protection, along with diodes placed in parallel with the controlled device to eliminate spikes.

Excitation dc power.   The dc power supply provides adequate power to drive the controlled device (motor). It must contain current-limiting circuitry to protect itself and final power drive from overload or shorting. Although not shown on the block diagram, dc power is also required to energize the electronic circuits within the various blocks.

ALTERNATING-CURRENT MOTORS

The alternating-current (ac) motor is similar in operation to the dc motor in that it has a field and a rotating armature. A single ac motor has a rotor that is an electromagnet. The field is provided by coils excited from an ac source. Rotation of the rotor is in phase with the field reversals directed by the frequency of the applied ac voltage. The rate of rotation is equal to the applied field ac frequency. Initial torque is low and the motor has poor speed control.

Alternating-Current Motor Conditioning

The alternating-current (ac) motor is an electromechanical device whose geometry and strength are important to control system hookups. Specifications will include rotor rotation speed, torque, and starting capabilities. Other specifications, such as ambient and working temperatures, power, voltage, and current, are significant. Functions required to condition the alternating-current (ac) motor for mechanical operation are as follows (see Figure 7–6):

1. PIO interface from the bus
2. Decoding
3. Buffering

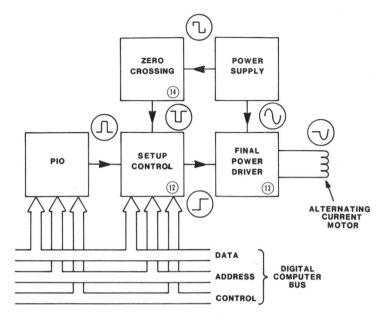

Figure 7–6 DIGITAL-TO-ANALOG CONVERSION: MOTOR (AC)

4. Latching
5. Zero-crossing detection
6. Power driving

Parallel input/output.   The purpose of the PIO is to interface the computer bus directly with the setup and control circuitry and/or the decoder. The stage must provide the "handshaking" between the setup and control circuitry and the computer by controlling data bus traffic direction and signaling the control service to tell it when to turn on or off.

Generally, the PIO is a single IC and is part of the computer's IC chip set. The PIO recognizes when it is to act/react by address lines from that bus. The PIO performs actions in response to instructions from the control bus. Information is taken or given by way of the data bus. Input to the PIO is a digital word from the computer. Output from the PIO is a digital word to the setup and control circuitry and/or decoder.

A more detailed explanation of the PIO is found in Chapter 8 in the section "The Microprocessor." PIO integrated circuits usually accompany the microprocessor chip and therefore are described along with the microprocessor.

Setup control for alternating-current application.   The purpose of the setup control is to provide circuitry to decode and to determine which device is to be set in operation. The setup control also provides circuits to buffer the computer bus and latch in the information as to which way to control the device either on or off. A final circuit accepts the output of the zero-crossing detector. This function enables the control device to turn on only when the zero-crossing detector acknowledges the go ahead. This action is required to eliminate a high current surge when the control device is turned on.

Final power drive.   The final power drive interfaces the setup control stage with the device to be controlled. This stage accepts a signal logic level of low power from the setup control stage and translates it into the power level required to operate the controlled device. The final power drive usually contains a power transistor switch, a triac, or a solid-state relay. In conjunction with functioned circuits, the final power drive must contain short-circuit protection for high-voltage spikes due to inductive surge. Circuits can be current limiters for power supply protection, along with diodes placed in parallel with the controlled device to eliminate spikes.

Zero-crossing detector.   The zero-crossing detector has one purpose. It provides an enable pulse each time the ac power supply current transitions through zero. This pulse is fed to the setup control for synchronization.

Alternating-current power supply.   The ac power supply energizes and provides current for the controlled device. It must contain enough capability to drive the number of controlled devices for which it is responsible. Further, it should provide current limiting to protect itself and the final power drive from burning up in the event of a short. Although not shown on the block diagram, dc power is required to energize the electronic circuits within the various blocks.

STEPPER MOTORS

The stepper motor is a multicoiled motor that rotates in increments to positions or steps, depending on which coil is energized. Each step is a final position unless further excitation takes place. Therefore, continuous rotation may be achieved by a pulse train in steps. Stepper motors may move in very small and precise increments according to its resolution.

Stepper Motor Conditioning

The stepper motor is an induction-type ac motor that depends on a rotor field induced by ac field coils. Each motor has a specific number of steps usually graded in a number of degrees (example 20° per step), at specific revolutions per minute (rpm). Other specifications include torque, temperature range, power, voltage, current, and mechanical hookups. Functions required to condition the stepper motor are as follows (see Figure 7–7):

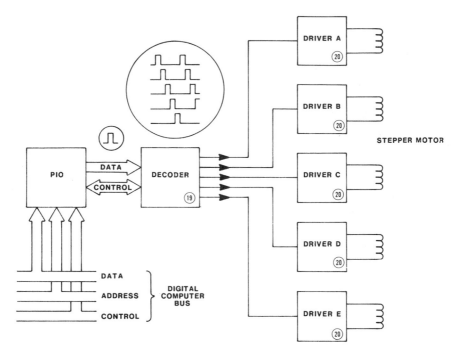

Figure 7-7 DIGITAL-TO-ANALOG CONVERSION: STEPPER MOTOR

1. PIO interface from the bus
2. Decoding
3. Power drivers A through E

Parallel input/output.   The purpose of the PIO is to interface the computer bus directly with the setup and control circuitry and/or the decoder. The stage must provide the "handshaking" between the decoder and the computer by controlling data bus traffic direction and signaling the control service to tell it when to turn on or off.

Generally, the PIO is a single IC and is part of the computer's IC chip set. The PIO recognizes when it is to act/react by address lines from that bus. The PIO performs actions in response to instructions from the control bus. Information is taken or given by way of the data bus. Input to the PIO is a digital word from the computer. Output from the PIO is a digital word to the decoder.

A more detailed explanation of the PIO is found in Chapter 8 in the section "The Microprocessor." PIO integrated circuits usually accompany the microprocessor chip and therefore are described along with the microprocessor.

Decoder.   The decoder includes electronics to latch in the data from the PIO and convert it into motor control logic. Generally, the decoder is a counter set up to ripple down pulses out to the various motor drivers when the motor is in a continuous rotating mode.

Drivers A through E.   There can be as many drivers as desired with a stepper motor, one for each coil. For the example shown in Figure 7–7, five drivers are illustrated (A through E). These drivers provide final drive circuits to translate the low power output of the decoder into high power drive to energize coils within the stepper motor. The number of drivers depend on the number of coils of the stepper motor and the resolution of the motor. The power driver usually consists of a power transistor with input logic.

Alternating-current power supply.   The ac power supply energizes and provides current for the controlled device. It must contain enough capability to drive the number of stepper motor drivers for which it is responsible. Further, it should provide current limiting to protect itself and the final power drive from burning up in the event of a short. Although not shown on the block diagram, dc power is also required to energize the electronic circuits within the various blocks.

ANALOG VOLTAGE

It is often necessary to output an analog voltage. The voltage could be used for a various number of applications, such as excitation power, reference, or as an analog comparison.

Analog Voltage Conditioning

An analog voltage output that represents a digital word must be conditioned. The word does represent some value. To condition a digital word to an analog output, the functions are as follows (see Figure 7–8):

1. Multiplying digital-to-analog (D/A) converter
2. Output amplifier

Multiplying digital-to-analog converter.   The multiplying D/A converter is a single-chip device with built-in latches to take data directly off the data bus, hold them and convert them into an analog voltage that is driven into an output amplifier. Multiplying D/A converters are available in various resolutions of 8, 12, 14, and 16 bits. Address lines from the bus can be used to enable and select the multiplying D/A converter. The D/A converter can be configured to create a programmable gain amplifier.

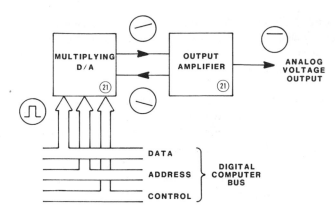

Figure 7–8 DIGITAL-TO-ANALOG CONVERSION: ANALOG VOLTAGE OUT-
PUT

Output amplifier. The purpose of the output amplifier is to amplify the
output of the D/A converter and place it at a usable range. The multiplying D/A
converter outputs a voltage through a resistor ladder network that can be amplified
by the output amplifier.

Excitation dc voltage. Note that there are no excitation power supplies
on the block diagram. There is a requirement for dc power to energize the electronics
within the various blocks.

8

Microprocessor Control

The *microprocessor* is a part of a microcomputer. It is a single integrated-circuit chip containing a part of the computer. The part it usually contains is called a central processing unit (CPU). Some microprocessor chips contain arithmetic units, control registers, and some memory. The microprocessor may indeed be the most thrilling and innovative electronic device ever conceived. The *microcomputer* is an extremely small (miniature) digital information-processing system consisting of integrated circuits and a microprocessor. The names ''microprocessor'' and ''microcomputer'' are used synonymously.

THE BASIC COMPUTER

The basic digital computer consists of three fundamental subsystems (see Figure 8–1). These are the input and output (I/O), the control processing unit (CPU), and the memory. They are all tied together by electrical control wiring and signal data buses.

The I/O unit is an input/output device that allows the operator to interface with the computer. It may input data and/or programs and achieve results from a readout or a printer. The memory unit stores the input, the program, and other data, and gives it up when told to do so. The central processing unit (CPU) performs mathematical and logical operations and controls the operation of all units. The data bus carries signal data between components. Control wiring carries commands to direct the system to function in its various modes.

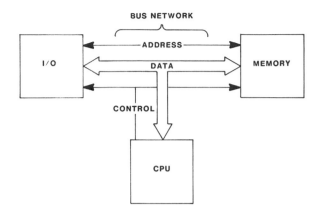

Figure 8–1 BASIC DIGITAL COMPUTER

SOME COMPUTER VERNACULAR

All computers have similar components and circuitry. Some standard terms can be applied to all computer systems.

Programming.   Instructions input to the computer to carry out a sequence of operations are called programs. These are applied by keyboard, tape, punched cards, and other methods. Programming may be directed to the processing unit or the memory.

Hardware.   The physical parts of the computer are called hardware. These parts may be major components such as the CPU, memory, I/O, and keyboard. Other hardware may be the entire package or wiring. Connectors and printed circuits also fall into this category.

Software.   Instructions to the computer are called software. The instruction may be in the form of keyboard inputs to memory or processor, magnetic tape, punched cards, paper tape, and so on. Programmed instructions are loaded into the computer memory called random-access memory (RAM).

Firmware.   These are programs that are fixed within the computer. These programs are called read-only memory (ROM).

ROM.   Read-only memory (ROM) is permanent programmed memory. It cannot be erased or modified. It may be called up (read out) but not written into. ROMs are not affected by shutoff of power to the computer.

RAM. Random-access memory is that memory which may be read out of, written into, erased, and reused. In other words, it may be modified by almost any means.

Volatile. The word "volatile" has to do with loss of memory. If memory is lost when power is turned off, the memory is termed volatile.

Nonvolatile. If memory is retained when power is turned off, the memory is termed nonvolatile.

External memory. Memory that is added to the basic computer and is interconnected to the CPU is external memory.

LANGUAGE

There are many language types used with the computer. Each has its place in computer operation and understanding. Operation and programming are accomplished in computer language. The lowest-level computer language is *machine language*. Machine language is a string of 0s and 1s used to trigger the gates and multivibrators that control the computer operation. Machine language consists of two sets of instructions that instruct the operation to be used and the address to be operated on.

Assembly language is an instruction from machine language translated into a mnemonic code such as CLR, which means "clear." It is a memory aid and allows the programmer to construct easily remembered instructions.

COBOL is an acronym meaning Common Business-Oriented Language. *FORTRAN* is another acronym, meaning FORmula TRANslation. Obviously, this is a mathematical-directed language. *BASIC* (Beginner's All-Purpose Symbolic Instruction Code), along with FORTRAN and COBOL, are procedure-directed languages. That is, they are used to direct the computer to perform some logical task rather than to deal with machine operation. Such operations as retrieving data from memory or loading data are basic and typical functions of these higher-level languages.

Although it would be meaningful to spend more time with language and the programming aspects of process control, these subjects are beyond the scope of this book. There are already many books published on the subject.

THE MICROPROCESSOR

The microprocessor is an integrated circuit (IC). It comes in typical IC form. This chip has made a permanent and almost awesome effect on the electronics industry. Its purpose is to process digital data inputs and to provide digital data outputs. The primary parts of the microprocessor are as follows:

1. The accumulator
2. Arithmetic logic unit
3. General register
4. Instruction register
5. Instruction decoder

The microprocessor is a controlling device. As discussed previously for the basic computer, the central processing unit (CPU) is the controlling unit for the basic computer. Its purpose is to perform mathematical and logical operations and control the operation of all the other units in the computer system. The microprocessor is the CPU of a microprocessor-based system. The functional characteristics (what it does) are based on internal architecture. Its architecture differs from one microprocessor to the other, but the basic functions are generally the same.

Pin Connections

The only method of getting in and/or out of the microprocessor is its terminal pins. A typical microprocessor has 40 such terminal pins. The pins, of course, are the hardware interface between the processing functions of the microprocessor and the other electronic peripheral parts of the control system. A list of pins and their functions is provided in Table 8–1.

TABLE 8–1   Pin Identification

Pin Name	Purpose	Direction In/Out
A_0–A_{15}	Address bus	Output
D_0–D_7	Data bus	Input/output
	Timing and control	
WR	Write	Output
DBIN	Data bus control	Output
INTE	Interrupt control	Output
INT	Interrupt control	Input
HLDA	Hold control	Output
HOLD	Hold control	Input
WAIT	Wait control	Output
READY	Wait control	Input
SYNC	Synchronize	Output
$\phi1$	Clock	Input
$\phi2$	Clock	Input
RESET	Reset	Input
	Power Supplies	
+12 v	Dc power	N/A
+5 V	Dc power	N/A
+5 V	Dc power	N/A
GND	Ground	N/A

SPECIAL MICROPROCESSOR INTERFACE CHIPS

To make a microprocessor operate like a computer, external circuitry must be utilized. The external circuitry allows the microcomputer to communicate with the memory, the external worlds, and special devices such as transducers. Some IC manufacturers make it easy to choose. They design peripheral ICs that do these tasks and are compatible with the microprocessor.

The *peripheral interface adapter* (PIA) is an interface chip designed to perform all input/output (I/O) functions, handshaking routines, and timing functions. It is also capable of performing serial-to-parallel or parallel-to-serial data conversions.

The *serial input/output* (SIO) is an IC that transmits data in a serial stream (one bit at a time). The SIO contains CPU input/output, internal control logic, interrupt control logic, discrete control, and two channels of serial output.

The *direct memory access* (DMA) provides direct access to the memory. By using this function it speeds up transfer between peripherals and memory by bypassing the CPU. Under usual conditions a transfer from memory requires an instruction to move the data from the memory to the accumulator first. DMA allows data to be placed directly from memory to peripheral without special accumulator instructions.

The *counter timer chip* (CTC) provides all timer and counter requirements for the microprocessor with which it is interfaced. The typical CTC has four channels, which may operate either as counters or timers at the same time. Other functions are CTC control and interrupt control.

The *parallel input/output* (PIO) chip is described in detail below.

PARALLEL INPUT/OUTPUT

The parallel input/output (PIO) is an IC that transmits data in a parallel stream. The PIO contains CPU input/output internal control logic, interrupt control, and one or two channels of parallel output.

A typical PIO chip is illustrated in Figure 8–2. This chip and the PIO description and functions are presented in the next several paragraphs. The PIO is utilized throughout this book as an interface in analog-to-digital (A/D) conversion and as an interface in digital-to-analog (D/A) conversion. Since it is a special microprocessor interface chip, we are covering it in some added detail.

Modes of Operations

The typical parallel input/output chip can be programmed in two modes of operation: the data-byte mode and the bit mode. The data-byte mode transfers a full digital word (8, 10, 12, 16 bits, etc.) to the data bus at the same time with handshaking. The bit mode transfers a single bit to the data bus while all others remain constant. The bit mode is done without handshaking. Handshaking, you

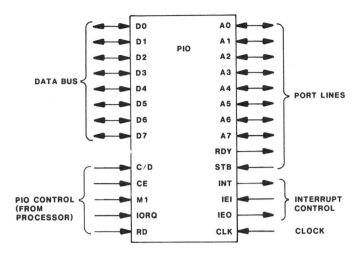

Figure 8–2 TYPICAL PIO CHIP

may recall, is a system of communication where one part of the microprocessor system transmits information and another part receives. The receiver then transmits acknowledgment back to the original sender. This is important in that a data ready signal is reset to continue operation. In more sophisticated systems, handshaking is available in both directions.

Within the two modes of operation the chip can be programmed to handle data in four ways. Byte input may program a word from the computer data bus to the PIO. Byte output may program a word from the PIO into the computer data bus. Byte input/out may program a word from the computer data bus into the PIO while a word is coming into the data bus from the PIO. Finally, a bit output/input may program single bits of information in both directions between computer data bus and the PIO.

Interrupt Control

An interrupt is the function of a microprocessor that allows the microprocessor to stop executing one program and begin executing another program from the memory. This function is a result of an outside response. Some of these interrupts may be an alarm, updating time, and input/output data. The central processing unit (CPU), a microprocessor, receives a request for interrupt. The interrupt control section handles all processor interrupt protocol.

The interrupt control is handled in a daisy-chain fashion which uses the interrupt enable input (IEI)/interrupt enable output (IEO) lines. When an interrupt occurs anywhere in the system, the IEI/IEO lines toggle and the interrupt (INT) line toggles the processor's INT line to signal the processor that an interrupt has occurred. When the processor learns that there is an interrupt, it interrogates the

PIO to find out where to look in the memory to determine what to do next and which peripheral caused the interrupt.

Handshaking Control

Handshaking is a system where parts of the microprocessor system send, receive, and acknowledge availability and receipt of data. The PIO has internal control logic that provides handshaking control by means of several functions. Internal control logic of the PIO provides the following functions:

1. Synchronizes port operation
2. Controls the port mode
3. Controls port addressing
4. Selects functions whether read or write
5. Issues appropriate commands to the port and interrupt control sections

Data Transfer Control

Two port locations establish control of the port, ready (RDY), and strobe (STB). The ready (RDY) pin, one of the port lines, is an output from the PIO that tells the peripheral that it is ready to transfer data. The strobe (STB) pin, also one of the port lines, allows the peripheral to strobe that particular port (port A in the example of Figure 8–2).

Clock Control

The microprocessor system clock (CLK) is connected to the PIO in order to synchronize the operations between the PIO and the microprocessor.

PIO Pinouts

Most pinouts on PIO ICs throughout industry are similar, if not the same. The Figure 8–2 pinouts are described below.

1. A_0–A_7: The 8-bit data input/output port whereby the peripheral device is connected.
2. RDY: The ready port.
3. STB: The strobe port.
4. D_0–D_7: The 8-bit data input/output port which is connected to the computer data bus.
5. C/D: PIO control that selects the type of data transfer to be performed. This control determines the direction of information on the data bus whether it be processor to PIO, or vice versa.

6. CE: Enables the PIO to operate. If there are more than one PIO in the system the CPU and software make the selection as to when each one operates.

7. RD: Enables the data transfer from the PIO to the processor.

8. IORQ: Input/output request is used in conjunction with the C/D, CE, and RD lines to transfer data to the processor.

9. M1: Machine cycle input from the processor which synchronizes the PIOs operation while fetching information from memory during the interrupt processor.

10. IEI: Interrupt input to the PIO from the system daisy chain.

11. IEO: Interrupt output from the PIO to the system daisy chain.

12. INT: Interrupt output from the PIO to the processor signaling the processor that an interrupt occurred.

13. CLK: Clock input to the PIO synchronizing the PIOs operation with the processor.

Timing Diagrams

Input/output (I/O) read/write timing cycles are shown in Figures 8–3 and 8–4. During each of the read or write cycles, the IORQ request line becomes active (goes low). If the operation is read, the RD line will react low. If the

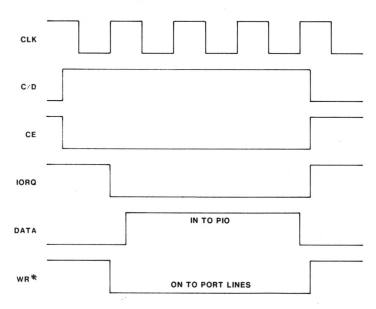

*INTERNAL SIGNAL GENERATED WITHIN THE PIO CHIP.

Figure 8–3 TIMING DIAGRAM WRITE CYCLE

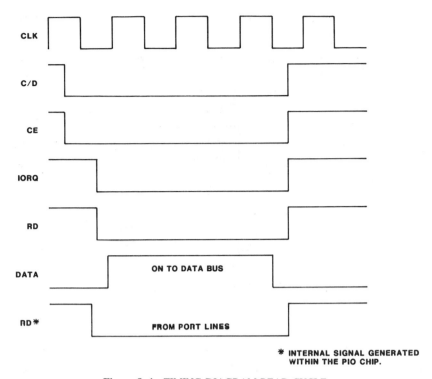

CLK

C/D

CE

IORQ

RD

DATA ON TO DATA BUS

RD* FROM PORT LINES

* INTERNAL SIGNAL GENERATED WITHIN THE PIO CHIP.

Figure 8-4 TIMING DIAGRAM READ CYCLE

operation is write, the WR line will react low. During both read and write, the data lines react high. C/D and CE control lines select and enable to start the process.

TYPICAL MICROPROCESSOR-CONTROLLED CLOSED LOOP

Figure 8–5 is a block diagram representing a closed-loop control system. The objective of this system is to control the position of a mechanism serviced by the servo valve. That mechanism can be part of any process that performs work, actuates, or in some other way makes motion. In this particular example the closed loop is a servomechanism. The servomechanism is direct digital controlled (DDC) by programmed software within the microprocessor. All driving signals to the servo valve are digital words which are passed through a multiplying D/A converter and amplification to exercise the current-activated servo valve. The servo valve is a high-response hydraulic valve which has a predictable and proportional relationship between an electrical current input and a hydraulic output. The servo valve applies an exact hydraulic output to the process actuator being manipulated. That hydraulic output is translated into some motion which is a mechanical link

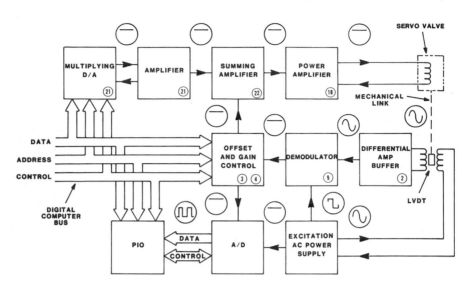

Figure 8-5 CLOSED-LOOP SYSTEM SERVO CYLINDER WITH LVDT FEED-
BACK

to the linear variable differential transformer (LVDT), a transducer. The LVDT
senses the position of the actuator driven by the servo valve. As the actuator
moves, it causes the LVDT to have an output that is opposite in polarity to the
signal originally received by the servo valve. This LVDT signal is a measure of
the position of the actuator. That signal, a feedback, is sensed by a differential
amplifier, demodulated, scaled, and converted by an A/D converter to a digital
word. It is then processed back to the computer by way of the PIO to be compared
against original driving information. The accuracy of the servomechanism's position
is dependent on the accuracy of the LVDT. The sensitivity of the servomechanism
is related to the gain of the driving signal and filtering throughout the loop. The
A/D converter instruments the system so that the computer may monitor system
performance. The driving function and the measurement function are analog. The
error detector and controlling functions are digital and, for the most part, are
part of the computer software. The multiplying D/A converter serves as the interface
from the bus to an analog that the process control can use.

In summary, the microprocessor contains the set point and software which
control the process to include error detection and integration. That software provides
digital words to transmit instructions through the D/A to the final control (servo
valve) and, in turn, the process being controlled. The transducer (an LVDT) is
the analog of the position (error) of the actuator driven by the servo valve. That
position feedback is converted by the A/D converter and fed to the microprocessor
through the PIO for comparison and evaluation to close the loop.

This entire process, at least what is represented in Figure 8–5, is a conditioning-type effort. The final control function is called the *forward path elements*. The measurement function is accomplished by the *feedback path elements*.

Forward Path Elements

The purpose of the forward path elements is to provide current to drive the servo valve by an amount established by the software program. Forward path elements include the following:

1. Multiplying D/A
2. Amplifier
3. Summing amplifier
4. Power amplifier
5. Servo valve

The *multiplying D/A* converter is a single-chip device with built-in latches used to remove data directly off of the data bus, hold them, and then convert them into an analog current. Address lines from the bus are used to enable and chip select the multiplying D/A converter.

The *amplifier* raises the level of the multiplying D/A to a usable range.

The *summing amplifier* sums the command voltage with the negative feedback voltage derived from the LVDT feedback path. This stage must have the correct amount of gain and filtering to establish proper system stability.

The *power amplifier* interfaces the summing amplifier with the servo valve. This stage translates a low power voltage from the summing amplifier into a high-level current to drive the servo valve.

Feedback Path Elements

The purpose of the feedback path elements is to provide position monitoring signals which reflect the position of the servomechanism. Feedback path elements include the following:

1. Linear variable differential transformer (LVDT)
2. Differential amplifier buffer
3. Demodulator
4. Offset and gain control
5. Analog-to-digital converter
6. Parallel input/output (PIO)

The *differential amplifier* and buffer measure the voltage across the secondary of the LVDT. The buffer must have the required impedance to load the LVDT properly.

The *demodulator* converts the analog into a dc voltage and determines the polarity of the dc voltage in relation to the phase of the analog. The demodulator acts as a phase-sensitive rectifier. It must be synchronized by ac power from the excitation ac power supply. A final purpose of the demodulator is to remove the inherent quadrature voltage or nulling voltage of the LVDT. This aids in aligning the mechanical null with the electrical null of the transducer.

The *offset and gain control* removes all the offset from the demodulator. It also adjusts the amplification of the demodulator voltage so that the output is properly scaled to the variable being measured. The offset and gain control is managed by the microprocessor directly from the bus by D/A converters and latches. Potentiometers are used for minor adjustment. This stage also contains filtering for stability.

The *A/D converter* converts the signal-conditioned analog to a digital word for further processing with the PIO.

The purpose of the *parallel input/output* is to interface the A/D converter directly to the computer bus. The PIO must also provide handshaking between the A/D converter and the microprocessor by controlling data bus traffic direction and signaling the microprocessor when a measured analog signal is to be released.

APPENDIX

Common Circuits for Block Diagrams in Text

The circuits illustrated in this appendix are common throughout Chapters 6, 7, and 8. They are typical and are representative of a sample circuit that could be utilized in the signal-conditioning process.

You will note that circuit values have been omitted (purposely). Each transducer requires different excitation and conditioning. Each must be treated independently. There is an assumption that circuits here will be understood by electronics people who are dealing with the transducers.

The circled numbers on the titles of the circuits shown here are those placed in the right lower corner of the block diagrams used in Chapters 6, 7, and 8.

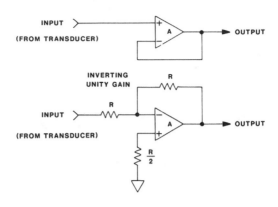

① SINGLE-ENDED BUFFER

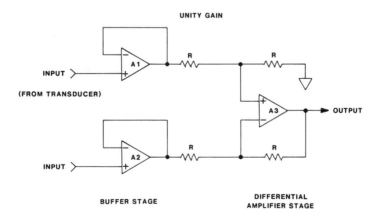

② DIFFERENTIAL AMPLIFIER AND BUFFER

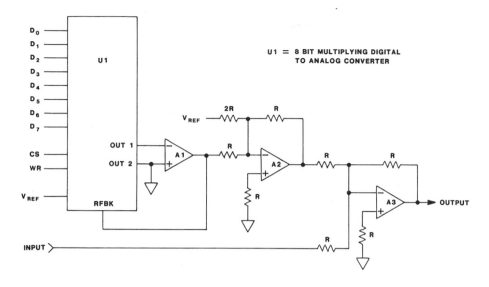

U1 = 8 BIT MULTIPLYING DIGITAL
TO ANALOG CONVERTER

③ **OFFSET CONTROL**

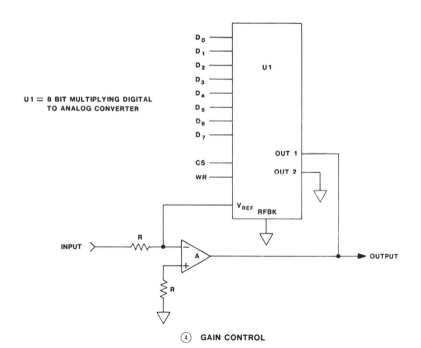

U1 = 8 BIT MULTIPLYING DIGITAL
TO ANALOG CONVERTER

④ **GAIN CONTROL**

R1 & R2   ADJUST COMPARATOR TRIP LEVEL
     C   ELIMINATES RINGING

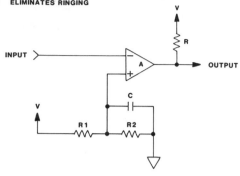

INPUT

V

R

A

+

OUTPUT

C

V

R1

R2

⑤ **BUFFER COMPARATOR**

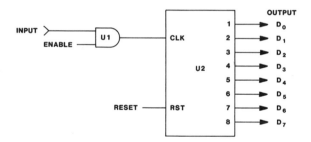

INPUT

ENABLE

U1

CLK

U2

RST

RESET

OUTPUT

1 → D_0
2 → D_1
3 → D_2
4 → D_3
5 → D_4
6 → D_5
7 → D_6
8 → D_7

U2 = BINARY COUNTER

⑥ **DECODER/PULSE COUNTER**

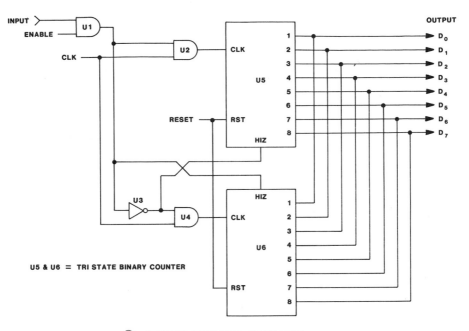

U5 & U6 = TRI STATE BINARY COUNTER

⑦   DECODER/FREQUENCY CONDITIONER

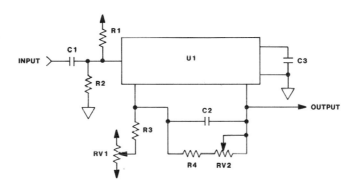

U1 = FREQUENCY TO VOLTAGE MICROCIRCUIT

C1 ELIMINATE DC
R1 & R2 SET UP THRESHOLD TO COMPARATOR
RV1 ADJUSTS OUTPUT AT LOW FREQUENCY
RV2 ADJUSTS OUTPUT AT HIGH FREQUENCY

⑧   FREQUENCY TO VOLTAGE CONVERTER

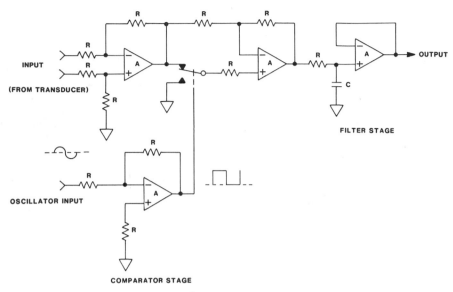

FILTER STAGE

COMPARATOR STAGE

⑨ DEMODULATOR

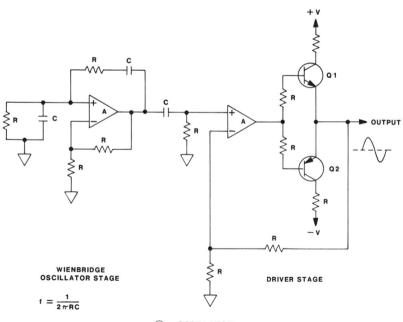

WIENBRIDGE
OSCILLATOR STAGE

$$f = \frac{1}{2\pi RC}$$

DRIVER STAGE

⑩ OSCILLATOR

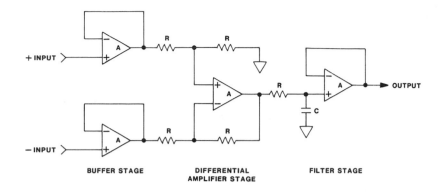

+ INPUT

− INPUT

BUFFER STAGE

DIFFERENTIAL
AMPLIFIER STAGE

FILTER STAGE

OUTPUT

⑪   **DIFFERENTIAL AMPLIFIER AND FILTER**

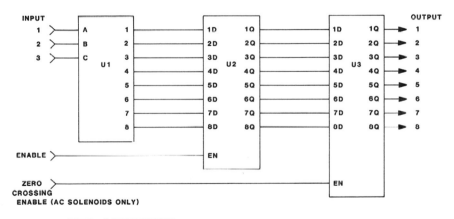

INPUT

OUTPUT

ENABLE

ZERO
CROSSING
ENABLE (AC SOLENOIDS ONLY)

U1 = 3 x 8 DEMULTIPLEXER
U2 & U3 = 8 BIT LATCH

⑫   **SET-UP CONTROL**

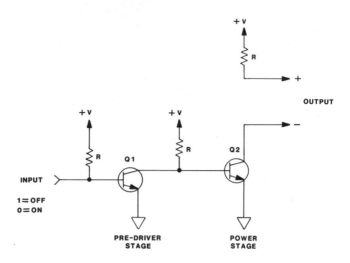

(13)   **FINAL POWER DRIVE**

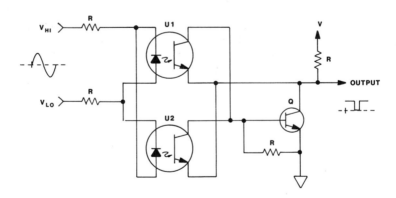

U1 & U2=OPTOISOLATOR

(14)   **ZERO CROSSING DETECTOR**

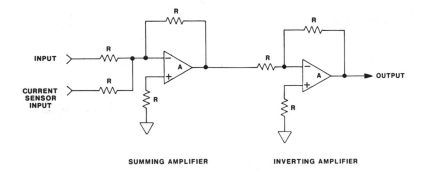

SUMMING AMPLIFIER   INVERTING AMPLIFIER

⑮   **PRE-AMPLIFIER**

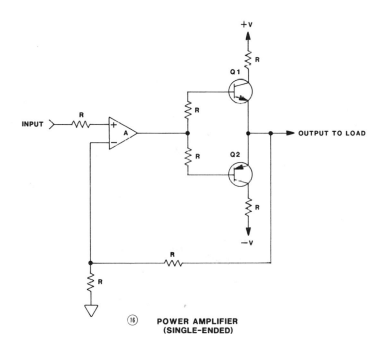

⑯   **POWER AMPLIFIER**
        (SINGLE-ENDED)

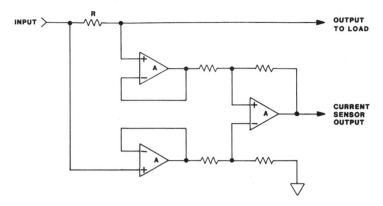

R NEEDS TO BE A RESISTANCE LOW ENOUGH TO
MINIMIZE ITS VOLTAGE DROP AS COMPARED TO THE LOAD.

⑰   **CURRENT SENSOR**

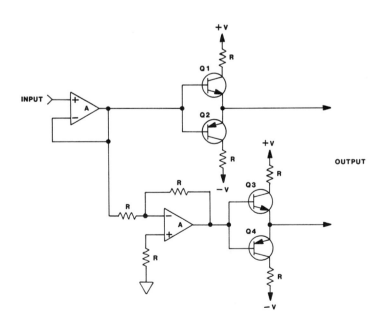

⑱   **POWER AMPLIFIER (DIFFERENTIAL ENDED)**

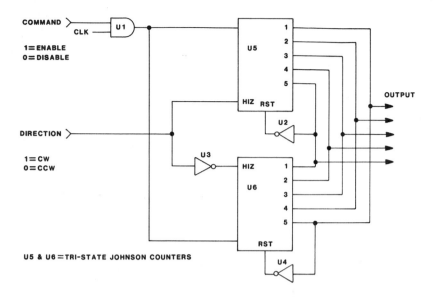

COMMAND

CLK — U1

1 = ENABLE
0 = DISABLE

U5

HIZ RST

U2

DIRECTION

1 = CW
0 = CCW

U3

HIZ

U6

OUTPUT

RST

U4

U5 & U6 = TRI-STATE JOHNSON COUNTERS

⑲   DECODER

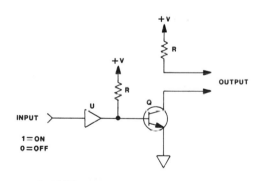

+V
R
OUTPUT

+V
R

U

INPUT

1 = ON
0 = OFF

Q

U = LOGIC -BUFFER

⑳   DRIVER

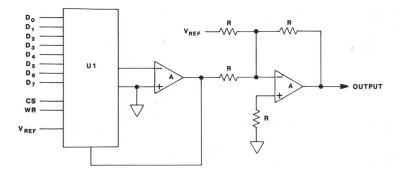

U1 = 8 BIT MULTIPLYING DIGITAL TO ANALOG CONVERTER

㉑   **OUTPUT AMPLIFIER**

INPUT 1 >——R——

INPUT 2 >——R——

R

R1

OUTPUT

OUTPUT = −(INPUT 1 + INPUT 2)

$$R1 = \frac{1}{\frac{1}{R} + \frac{1}{R} + \frac{1}{R}} = \frac{R}{3}$$

㉒   **SUMMING AMPLIFIER**

Index